GLENCOE
MATHEMATICS

D1268441

Noteables™

Interactive Study Notebook

with FOLDABLES™

Algebra 2

Contributing Author
Dinah Zike

FOLDABLES™

Consultant
Douglas Fisher, PhD
Director of Professional Development
San Diego State University
San Diego, CA

McGraw Hill Glencoe

New York, New York   Columbus, Ohio   Chicago, Illinois   Peoria, Illinois   Woodland Hills, California

Mc Graw Hill **Glencoe**

The *McGraw·Hill* Companies

Send all inquiries to:
The McGraw-Hill Companies
8787 Orion Place
Columbus, OH 43240-4027

Aquinas College Library
1607 Robinson Rd. SE
Grand Rapids, MI 49506

ISBN: 0-07-868209-6

Algebra 2 (Student Edition)
Noteables™: Interactive Study Notebook with Foldables™

1 2 3 4 5 6 7 8 9 10 047 09 08 07 06 05 04

Contents

Organizing Your Foldables

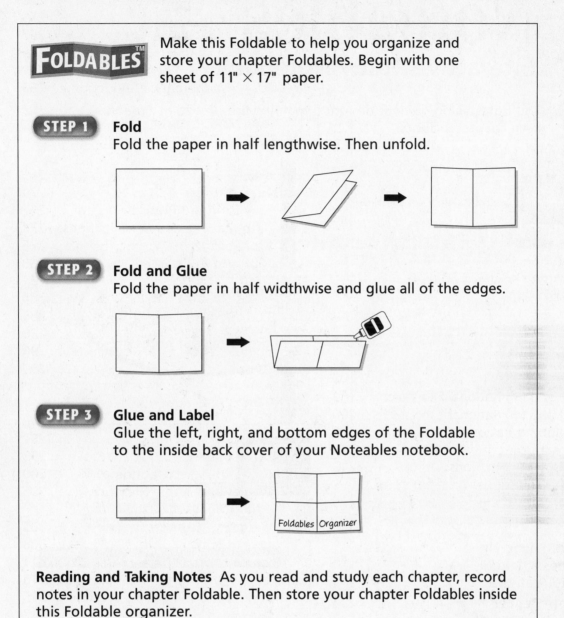

FOLDABLES™ Make this Foldable to help you organize and store your chapter Foldables. Begin with one sheet of 11" × 17" paper.

STEP 1 Fold
Fold the paper in half lengthwise. Then unfold.

STEP 2 Fold and Glue
Fold the paper in half widthwise and glue all of the edges.

STEP 3 Glue and Label
Glue the left, right, and bottom edges of the Foldable to the inside back cover of your Noteables notebook.

Foldables | Organizer

Reading and Taking Notes As you read and study each chapter, record notes in your chapter Foldable. Then store your chapter Foldables inside this Foldable organizer.

Using Your
Noteables™ with FOLDABLES™
Interactive Study Notebook

This note-taking guide is designed to help you succeed in *Algebra 2*.
Each chapter includes:

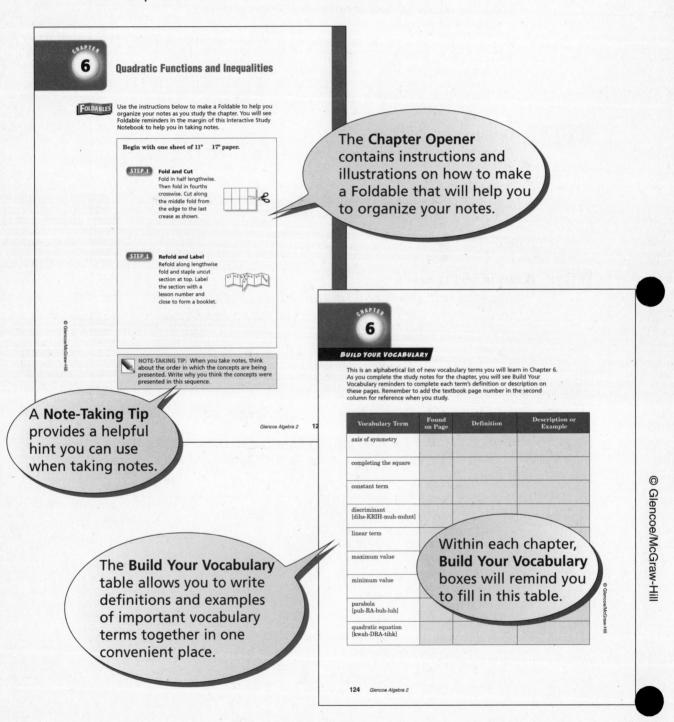

The **Chapter Opener** contains instructions and illustrations on how to make a Foldable that will help you to organize your notes.

A **Note-Taking Tip** provides a helpful hint you can use when taking notes.

The **Build Your Vocabulary** table allows you to write definitions and examples of important vocabulary terms together in one convenient place.

Within each chapter, **Build Your Vocabulary** boxes will remind you to fill in this table.

© Glencoe/McGraw-Hill

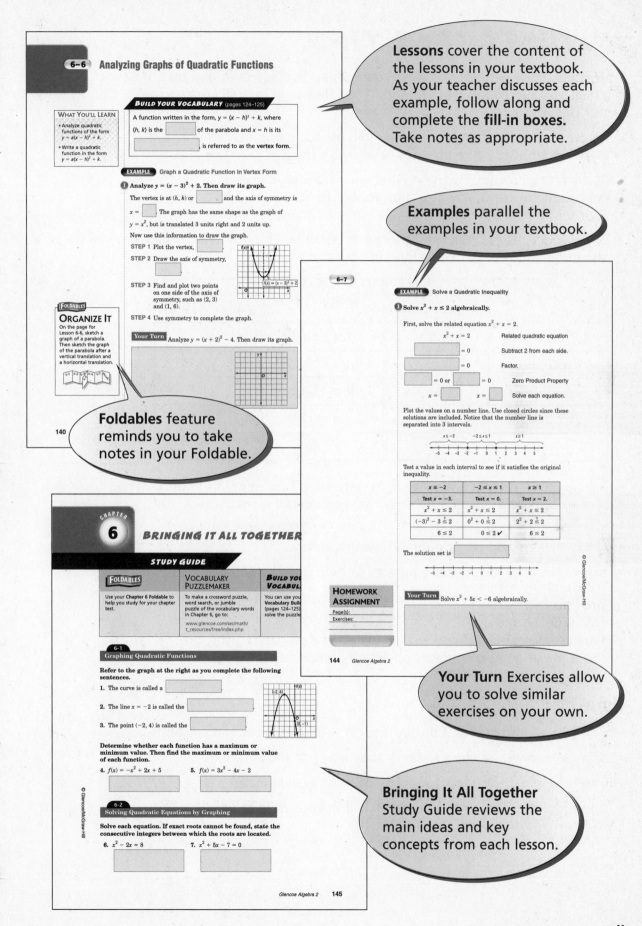

6-6 Analyzing Graphs of Quadratic Functions

WHAT YOU'LL LEARN
• Analyze quadratic functions of the form $y = a(x - h)^2 + k$.
• Write a quadratic function in the form $y = a(x - h)^2 + k$.

BUILD YOUR VOCABULARY (pages 124–125)

A function written in the form, $y = (x - h)^2 + k$, where (h, k) is the ⬚ of the parabola and $x = h$ is its ⬚, is referred to as the **vertex form**.

EXAMPLE Graph a Quadratic Function in Vertex Form

① Analyze $y = (x - 3)^2 + 2$. Then draw its graph.

The vertex is at (h, k) or ⬚ and the axis of symmetry is $x = $ ⬚. The graph has the same shape as the graph of $y = x^2$, but is translated 3 units right and 2 units up.

Now use this information to draw the graph.

STEP 1 Plot the vertex, ⬚.

STEP 2 Draw the axis of symmetry, ⬚.

STEP 3 Find and plot two points on one side of the axis of symmetry, such as (2, 3) and (1, 6).

STEP 4 Use symmetry to complete the graph.

FOLDABLES
ORGANIZE IT
On the page for Lesson 6-6, sketch a graph of a parabola. Then sketch the graph of the parabola after a vertical translation and a horizontal translation.

Your Turn Analyze $y = (x + 2)^2 - 4$. Then draw its graph.

140

Foldables feature reminds you to take notes in your Foldable.

Lessons cover the content of the lessons in your textbook. As your teacher discusses each example, follow along and complete the **fill-in boxes.** Take notes as appropriate.

Examples parallel the examples in your textbook.

6-7

EXAMPLE Solve a Quadratic Inequality

③ Solve $x^2 + x \leq 2$ algebraically.

First, solve the related equation $x^2 + x = 2$.

$x^2 + x = 2$		Related quadratic equation
⬚ $= 0$		Subtract 2 from each side.
⬚ $= 0$		Factor.
⬚ $= 0$ or ⬚ $= 0$		Zero Product Property
$x = $ ⬚	$x = $ ⬚	Solve each equation.

Plot the values on a number line. Use closed circles since these solutions are included. Notice that the number line is separated into 3 intervals.

Test a value in each interval to see if it satisfies the original inequality.

$x \leq -2$	$-2 \leq x \leq 1$	$x \geq 1$
Test $x = -3$.	Test $x = 0$.	Test $x = 2$.
$x^2 + x \leq 2$	$x^2 + x \leq 2$	$x^2 + x \leq 2$
$(-3)^2 - 3 \overset{?}{\leq} 2$	$0^2 + 0 \overset{?}{\leq} 2$	$2^2 + 2 \overset{?}{\leq} 2$
$6 \leq 2$	$0 \leq 2$ ✔	$6 \leq 2$

The solution set is ⬚.

Your Turn Solve $x^2 + 5x < -6$ algebraically.

HOMEWORK ASSIGNMENT
Page(s):
Exercises:

144   Glencoe Algebra 2

Your Turn Exercises allow you to solve similar exercises on your own.

CHAPTER 6

BRINGING IT ALL TOGETHER

STUDY GUIDE

FOLDABLES
Use your Chapter 6 Foldable to help you study for your chapter test.

VOCABULARY PUZZLEMAKER
To make a crossword puzzle, word search, or jumble puzzle of the vocabulary words in Chapter 6, go to:
www.glencoe.com/sec/math/ t_resources/free/index.php

BUILD YOUR VOCABULARY
You can use your Vocabulary Builder (pages 124–125) to solve the puzzles.

6-1
Graphing Quadratic Functions

Refer to the graph at the right as you complete the following sentences.

1. The curve is called a ⬚.

2. The line $x = -2$ is called the ⬚.

3. The point $(-2, 4)$ is called the ⬚.

Determine whether each function has a maximum or minimum value. Then find the maximum or minimum value of each function.

4. $f(x) = -x^2 + 2x + 5$ ⬚

5. $f(x) = 3x^2 - 4x - 2$ ⬚

6-2
Solving Quadratic Equations by Graphing

Solve each equation. If exact roots cannot be found, state the consecutive integers between which the roots are located.

6. $x^2 - 2x = 8$ ⬚

7. $x^2 + 5x - 7 = 0$ ⬚

Glencoe Algebra 2   145

Bringing It All Together Study Guide reviews the main ideas and key concepts from each lesson.

NOTE-TAKING TIPS

Your notes are a reminder of what you learned in class. Taking good notes can help you succeed in mathematics. The following tips will help you take better classroom notes.

- Before class, ask what your teacher will be discussing in class. Review mentally what you already know about the concept.

- Be an active listener. Focus on what your teacher is saying. Listen for important concepts. Pay attention to words, examples, and/or diagrams your teacher emphasizes.

- Write your notes as clear and concise as possible. The following symbols and abbreviations may be helpful in your note-taking.

Word or Phrase	Symbol or Abbreviation	Word or Phrase	Symbol or Abbreviation
for example	e.g.	not equal	$\neq$
such as	i.e.	approximately	$\approx$
with	w/	therefore	$\therefore$
without	w/o	versus	vs
and	+	angle	$\angle$

- Use a symbol such as a star ($\star$) or an asterisk (*) to emphasis important concepts. Place a question mark (?) next to anything that you do not understand.

- Ask questions and participate in class discussion.

- Draw and label pictures or diagrams to help clarify a concept.

- When working out an example, write what you are doing to solve the problem next to each step. Be sure to use your own words.

- Review your notes as soon as possible after class. During this time, organize and summarize new concepts and clarify misunderstandings.

Note-Taking Don'ts

- **Don't** write every word. Concentrate on the main ideas and concepts.
- **Don't** use someone else's notes as they may not make sense.
- **Don't** doodle. It distracts you from listening actively.
- **Don't** lose focus or you will become lost in your note-taking.

Solving Equations and Inequalities

 Use the instructions below to make a Foldable to help you organize your notes as you study the chapter. You will see Foldable reminders in the margin of this Interactive Study Notebook to help you in taking notes.

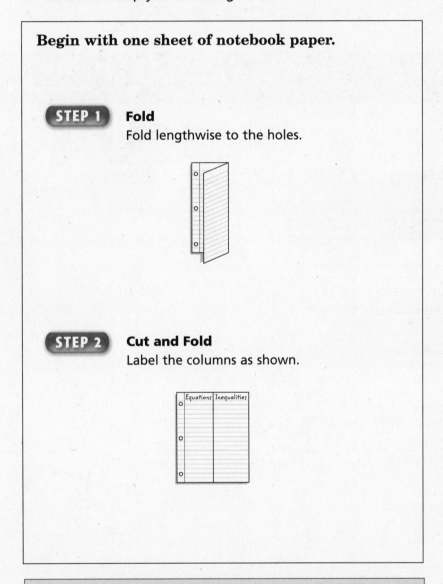

Begin with one sheet of notebook paper.

STEP 1 **Fold**
Fold lengthwise to the holes.

STEP 2 **Cut and Fold**
Label the columns as shown.

Equations	Inequalities

 NOTE-TAKING TIP: When you take notes, it is often a good idea to use symbols to emphasize important concepts.

This is an alphabetical list of new vocabulary terms you will learn in Chapter 1. As you complete the study notes for the chapter, you will see Build Your Vocabulary reminders to complete each term's definition or description on these pages. Remember to add the textbook page number in the second column for reference when you study.

Vocabulary Term	Found on Page	Definition	Description or Example
absolute value			
algebraic expression			
Associative Property [uh-SOH-shee-uh-tihv]			
Commutative Property [kuh-MYOO-tuh-tihv]			
compound inequality			
Distributive Property [dih-STRIH-byuh-tihv]			
empty set			
Identity Property			
intersection			
Inverse Property			

Vocabulary Term	Found on Page	Definition	Description or Example
irrational numbers			
open sentence			
rational numbers			
Reflexive Property			
set-builder notation			
Substitution Property			
Symmetric Property [suh-MEH-trihk]			
Transitive Property			
Trichotomy Property [try-KAH-tuh-mee]			
union			

1-1 Expressions and Formulas

WHAT YOU'LL LEARN

- Use the order of operations to evaluate expressions.

- Use formulas.

BUILD YOUR VOCABULARY (page 2)

Expressions that contain at least one variable are called **algebraic expressions**.

EXAMPLE Simplify an Expression

1 Find the value of $[384 - 3(7 - 2)^3] \div 3$.

$[384 - 3(7 - 2)^3] \div 3$

$= [384 - 3(5)^3] \div 3$ Subtract 2 from 7.

$= [384 - 3(\boxed{})] \div 3$ Then cube 5.

$= (384 - 375) \div 3$ Multiply 125 by 3.

$= \boxed{} \div 3$ Subtract 375 from 384.

$= \boxed{}$ Divide 9 by 3.

Your Turn Find the value of $[26 - 2(7 - 5)^2] \div 2$.

KEY CONCEPT

Order of Operations

1. Evaluate expressions inside grouping symbols, such as parentheses, ( ), brackets, [ ], braces, { }, and fraction bars, as in $\frac{5 + 7}{2}$.

2. Evaluate all powers.

3. Do all multiplications and/or divisions from left to right.

4. Do all additions and/or subtractions from left to right.

FOLDABLES Write the order of operations in the Equations column of your Foldable.

EXAMPLE Evaluate an Expression

2 Evaluate $s - t(s^2 - t)$ if $s = 2$ and $t = 3.4$.

$s - t(s^2 - t) = 2 - 3.4(2^2 - 3.4)$ Replace s with 2 and t with 3.4.

$= 2 - 3.4(4 - 3.4)$ Find 2^2.

$= 2 - 3.4\left(\boxed{}\right)$ Subtract 3.4 from 4.

$= 2 - \boxed{}$ Multiply 3.4 and 0.6.

$= \boxed{}$ Subtract 2.04 from 2.

Your Turn Evaluate $x - y^2(x + 5)$ if $x = 2$ and $y = 4$.

EXAMPLE Expression Containing a Fraction Bar

3 Evaluate $\dfrac{8xy + z^3}{y^2 + 5}$ if $x = 5$, $y = -2$, and $z = -1$.

$$\dfrac{8xy + z^3}{y^2 + 5} = \dfrac{8(5)(-2) + (-1)^3}{(-2)^2 + 5}$$

$x = 5$, $y = -2$, and $z = -1$

$$= \dfrac{\boxed{}(-2) + (-1)}{\boxed{} + 5}$$

Evaluate the numerator and the denominator separately.

$$= \dfrac{\boxed{}-1}{4 + 5}$$

Multiply 40 by –2.

$$= \dfrac{-81}{9} \text{ or } \boxed{}$$

Simplify the numerator and the denominator. Then divide.

Your Turn Evaluate $\dfrac{3ab + c^2}{b - c}$ if $a = 5$, $b = 2$, and $c = 4$.

EXAMPLE Use a Formula

4 Find the area of a trapezoid with base lengths of 13 meters and 25 meters and a height of 8 meters.

$$A = \tfrac{1}{2}h(b_1 + b_2)$$ Area of a trapezoid

$$= \tfrac{1}{2}(8)(13 + 25)$$ Replace h with 8, b_1 with 13, and b_2 with 25.

$$= \boxed{}$$ Add 13 and 25.

$$= \boxed{}$$ Multiply 8 by $\tfrac{1}{2}$.

$$= \boxed{}$$ Multiply 4 and 38.

The area of the trapezoid is $\boxed{}$ square meters.

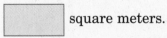 **Your Turn** The formula for the volume V of a pyramid is $V = \tfrac{1}{3}Bh$, where B is the area of the base and h is the height of the pyramid. Find the volume of the pyramid shown.

WRITE IT

Why is it important to follow the order of operations when evaluating expressions?

_____

_____

_____

_____

_____

_____

HOMEWORK ASSIGNMENT

Page(s): _____

Exercises: _____

Properties of Real Numbers

WHAT YOU'LL LEARN

- Classify real numbers.

- Use the properties of real numbers to evaluate expressions.

KEY CONCEPTS

Real Numbers

Rational Numbers A rational number can be expressed as a ratio $\frac{m}{n}$, where m and n are integers and n is not zero. The decimal form of a rational number is either a terminating or repeating decimal.

Irrational Numbers A real number that is not rational is irrational. The decimal form of an irrational number neither terminates nor repeats.

EXAMPLE Classify Numbers

1 Name the sets of numbers to which each number belongs.

a. $-\dfrac{2}{3}$

b. $9.99\overline{9}\ldots$

The bar over the 9 indicates that those digits repeat forever.

c. $\sqrt{6}$

$\sqrt{6}$ lies between 2 and 3 so it is not a whole number.

d. $\sqrt{100}$

$\sqrt{100} = 10$

EXAMPLE Identify Properties of Real Numbers

2 Name the property illustrated by each equation.

a. $(-8 + 8) + 15 = 0 + 15$

The Property says that a number plus its opposite is 0.

b. $5(8 - 6) = 5(8) - 5(6)$

The  Property says that you multiply each term within the parentheses by the first number.

KEY CONCEPT

Real Number Properties
For any real numbers a, b, and c:

Commuative
$a + b = b + a$ and
$a \cdot b = b \cdot a$

Associative
$(a + b) + c = a + (b + c)$
and $(a \cdot b) \cdot c = a \cdot (b \cdot c)$

Identity
$a + 0 = a = 0 + a$ and
$a \cdot 1 = a = 1 \cdot a$

Inverse
$a + (-a) = 0 = (-a) + a$
If $a \neq 0$, then $a \cdot \frac{1}{a} = 1$
$= \frac{1}{a} \cdot a$.

Distributive
$a(b + c) = ab + ac$ and
$(b + c)a = ba + ca$

HOMEWORK ASSIGNMENT

Page(s): _____

Exercises: _____

_____

_____

Your Turn Name the sets of numbers to which each number belongs.

a. $\frac{3}{5}$ [_____]

b. $-2.\overline{52}$ [_____]

c. $\sqrt{5}$ [_____]

d. $\sqrt{121}$ [_____]

Name the property illustrated by each equation.

e. $3 + 0 = 3$ [_____]

f. $5 \cdot \frac{1}{5} = 1$ [_____]

EXAMPLE Additive and Multiplicative Inverses

3 Identify the additive and multiplicative inverse of -7.

Since $-7 + 7 = 0$, the additive inverse is [____]. Since

$(-7)\left(-\frac{1}{7}\right) = 1$, the multiplicative inverse is [____].

Your Turn Identify the additive inverse and multiplicative inverse for each number.

a. 5 [_____]

b. $-\frac{2}{3}$ [_____]

EXAMPLE Use the Distributive Property to Solve a Problem

4 POSTAGE Audrey went to the post office and bought eight 34-cent stamps and eight 21-cent postcard stamps. How much did Audrey spend altogether on stamps?

To find the total amount spent on stamps, multiply the price of each type of stamp by 8 and then add.

$S = 8(0.34) + 8(0.21)$

$= [\quad] + [\quad]$ or $[\quad]$

So, Audrey spent [\quad] on stamps.

Your Turn Joel went to the grocery store and bought 3 plain chocolate candy bars for $0.69 each and 3 chocolate-peanut butter candy bars for $0.79 each. How much did Joel spend altogether on candy bars?

[_____]

EXAMPLE Verbal to Algebraic Expression

WHAT YOU'LL LEARN

- Translate verbal expressions into algebraic expressions and equations, and vice versa.

- Solve equations using the properties of equality.

1 Write an algebraic expression to represent

a. 3 more than a number

b. 6 times the cube of a number

Your Turn Write an algebraic expression to represent each verbal expression.

a. 2 less than the cube of a number

b. 10 decreased by the product of a number and 2

BUILD YOUR VOCABULARY (page 3)

A mathematical sentence containing one or more

[   ] is called an **open sentence.**

EXAMPLE Algebraic to Verbal Sentence

2 Write a verbal sentence to represent each equation.

a. $14 + 9 = 23$ The [   ] of 14 and 9 is 23.

b. $6 = -5 + x$ [   ] is equal to [   ] plus a number.

Your Turn Write a verbal sentence to represent each equation.

a. $5 = 2 + x$

b. $3a + 2 = 11$

KEY CONCEPT

Properties of Equality

Reflexive For any real number a, $a = a$.

Symmetric For all real numbers a and b, if $a = b$, then $b = a$.

Transitive For all real numbers a, b, and c, if $a = b$ and $b = c$, then $a = c$.

Substitution If $a = b$, then a may be replaced by b and b may be replaced by a.

EXAMPLE Identify Properties of Equality

3 Name the property illustrated by each statement.

a. If $xy = 28$ and $x = 7$, then $7y = 28$.

b. $a - 2.03 = a - 2.03$

Your Turn Name the property illustrated by each statement.

a. If $x + 4 = 3$, then $3 = x + 4$.

b. If $3 = x$ and $x = y$, then $3 = y$.

EXAMPLE Solve One-Step Equations

4 Solve each equation.

a. $s - 5.48 = 0.02$

$s - 5.48 = 0.02$	Original equation
$s - 5.48 +$ ☐ $= 0.02 +$ ☐	Add 5.48 to each side.
$s =$ ☐	Simplify.

b. $18 = \frac{1}{2}t$

$18 = \frac{1}{2}t$	Original equation
☐ $18 = \frac{1}{2}t$ ☐	Multiply each side by the multiplicative inverse of $\frac{1}{2}$.
☐ $= t$	Simplify.

Your Turn Solve each equation.

a. $x + 5 = 3$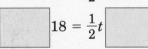

b. $\frac{2}{3}x = 10$

EXAMPLE Solve a Multi-Step Equation

⑤ Solve $53 = 3(y - 2) - 2(3y - 1)$.

$53 = 3(y - 2) - 2(3y - 1)$	Original equation
$53 = 3y - 6 - 6y + 2$	Distributive and Substitution Properties
$53 = \boxed{}$	Commutative, Distributive, and Substitution Properties
$\boxed{} = \boxed{}$	Addition and Substitution Properties
$\boxed{} = y$	Division and Substitution Properties

Your Turn Solve $25 = 3(2x + 2) - 5(2x + 1)$.

EXAMPLE Writing an Equation

⑥ **HOME IMPROVEMENT** Carl wants to replace 5 windows in his home. His neighbor Will is a carpenter and he has agreed to help install them for $250. If Carl has budgeted $1000 for the total cost, what is the maximum amount he can spend on each window?

Let c represent the cost of each window. Write and solve an equation to find the value of c.

Number of windows	times	cost of each window	plus	installation	equals	total cost.
5	·	c	+	250	=	1000

$5c + 250 = 1000$	Original equation
$5c + 250 - \boxed{} = 1000 - \boxed{}$	Subtract.
$5c = 750$	Simplify.
$c = \boxed{}$	Divide each side by 5.

Carl can afford to spend $\boxed{}$ on each window.

Your Turn Kelly wants to repair the siding on her house. Her contractor will charge her $300 plus $150 per square foot of siding. How much siding can she repair for $1500?

Solving Absolute Value Equations

WHAT YOU'LL LEARN

- Evaluate expressions involving absolute values.

- Solve absolute value equations.

BUILD YOUR VOCABULARY (page 2)

The **absolute value** of a number is its distance from ☐ on the number line.

The solution set for an equation that has no solution is the **empty set**, symbolized by ☐ or ☐ .

EXAMPLE Evaluate an Expression with Absolute Value

1 Evaluate $2.7 + |6 - 2x|$ if $x = 4$.

$2.7 + |6 - 2x| = 2.7 + |6 - 2(4)|$ Replace x with 4.

$= $ ☐   Simplify $-2(4)$ first.

$= $ ☐   Subtract 8 from 6.

$= $ ☐ $|-2| = 2$

$= $ ☐   Add.

Your Turn Evaluate $2.3 - |3y - 10|$ if $y = -2$.

REVIEW IT

What is the difference between an algebraic expression and an equation?
(Lessons 1-1, 1-3)

_____

_____

_____

_____

_____

_____

EXAMPLE Solve an Absolute Value Equation

2 Solve $|y + 3| = 8$.

Case 1 $a = b$ **Case 2** $a = -b$

$y + 3 = 8$ $y + 3 = -8$

$y + 3 - $ ☐ $= 8 - $ ☐ $y + 3 - $ ☐ $= -8 - $ ☐

$y = $ ☐ $y = $ ☐

The solutions are ☐ .

Thus, the solution set is ☐ .

EXAMPLE No Solution

3 Solve $|6 - 4t| + 5 = 0$.

$|6 - 4t| + 5 = 0$ Original equation

$|6 - 4t| = \boxed{}$ Subtract $\boxed{}$ from each side.

This sentence is *never* true. So, the solution set is $\boxed{}$.

EXAMPLE One Solution

4 Solve $|8 + y| = 2y - 3$. Check your solutions.

Case 1 $a = b$
$$8 + y = 2y - 3$$
$$\boxed{} = \boxed{}$$
$$\boxed{} = y$$

Case 2 $a = -b$
$$8 + y = -(2y - 3)$$
$$8 + y = -2y + 3$$
$$\boxed{} = 3$$
$$\boxed{} = \boxed{}$$
$$y = \boxed{}$$

There appear to be two solutions.

Check:

$|8 + y| = 2y - 3$

$\left|8 + \boxed{}\right| = 2\left(\boxed{}\right) - 3$

$\boxed{} \stackrel{?}{=} 19$

$19 = \boxed{}$

$|8 + y| = 2y - 3$

$\left|8 + \boxed{}\right| \stackrel{?}{=} 2\left(\boxed{}\right) - 3$

$\left|\dfrac{9}{13}\right| \stackrel{?}{=} \boxed{} - 3$

$\dfrac{9}{13} = \boxed{}$

Since $\dfrac{19}{3} \neq -\dfrac{19}{3}$, the only solution is $\boxed{}$.

Your Turn Solve each equation. Check your solutions.

a. $|2x + 5| = 15$

$\boxed{}$

b. $\left|5x - \dfrac{2}{3}\right| + 7 = 0$

$\boxed{}$

c. $|3x - 5| = -4x + 37$

$\boxed{}$

Solving Inequalities

© Glencoe/McGraw-Hill

WHAT YOU'LL LEARN

- Solve inequalities.
- Solve real-world problems involving inequalities.

BUILD YOUR VOCABULARY (page 3)

The **Trichotomy Property** says that, for any two real numbers, a and b, exactly [   ] of the following statements is true.

$$a < b \qquad a = b \qquad a > b$$

The [   ] of an inequality can be expressed by using **set-builder notation**, for example, $\{x \mid x > 9\}$.

KEY CONCEPTS

Properties of Inequality

Addition Property of Inequality For any real numbers a, b, and c:

If $a > b$, then
$a + c > b + c$.
If $a < b$, then
$a + c < b + c$.

Subtraction Property of Inequality For any real numbers a, b, and c:

If $a > b$, then
$a - c > b - c$.
If $a < b$, then
$a - c < b - c$.

EXAMPLE Solve an Inequality Using Addition or Subtraction

1 Solve $4y - 3 < 5y + 2$. Graph the solution set on a number line.

$4y - 3 < 5y + 2$ — Original inequality

$4y - 3 - \boxed{} < 5y + 2 - \boxed{}$ — Subtract $4y$ from each side.

$\boxed{} < \boxed{}$ — Simplify.

$\boxed{} < \boxed{}$ — Subtract from each side.

$\boxed{} < y \text{ or } y > \boxed{}$ — Simplify.

Any real number greater than -5 is a solution of this inequality.

A circle means that this point is **not** included in the solution set.

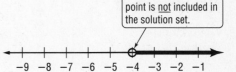

Your Turn Solve $6x - 2 < 5x + 7$. Graph the solution on a number line.

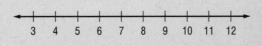

EXAMPLE Solve an Inequality Using Multiplication or Division

KEY CONCEPTS

Properties of Inequality

Multiplication Property of Inequality For any real numbers a, b, and c, where c is positive:

if $a > b$, then $ac > bc$.

if $a < b$, then $ac < bc$.

c is negative:

if $a > b$, then $ac < bc$.

if $a < b$, then $ac > bc$.

Division Property of Inequality For any real numbers a, b, and c, where c is positive:

if $a > b$, then $\frac{a}{c} > \frac{b}{c}$.

if $a < b$, then $\frac{a}{c} < \frac{b}{c}$.

c is negative:

if $a > b$, then $\frac{a}{c} < \frac{b}{c}$.

if $a < b$, then $\frac{a}{c} > \frac{b}{c}$.

2 Solve $12 \geq -0.3p$. Graph the solution set on a number line.

$$12 \geq -0.3p \qquad \text{Original inequality}$$

$$\frac{12}{\boxed{}} \leq \frac{-0.3p}{\boxed{}} \qquad \text{Divide each side by } \boxed{}, \text{ reversing the inequality symbol.}$$

$$\boxed{} \leq p \qquad \text{Simplify.}$$

$$p \geq \boxed{} \qquad \text{Rewrite with } p \text{ first.}$$

The solution set is $\{p \,|\, p \geq -40\}$.

A dot means that this point is included in the solution set.

$$-42 \quad -41 \quad -40 \quad -39 \quad -38 \quad -37 \quad -36$$

EXAMPLE Solve a Multi-Step Inequality

3 Solve $-x > \frac{x-7}{2}$. Graph the solution set on a number line.

$$-x > \frac{x-7}{2} \qquad \text{Original inequality}$$

$$-2x > x - 7 \qquad \text{Multiply each side by 2.}$$

$$\boxed{} > \boxed{} \qquad \text{Add } -x \text{ to each side.}$$

$$x < \boxed{} \qquad \text{Divide each side by } -3, \text{ reversing the inequality symbol.}$$

The solution set is $\left(-\infty, \frac{7}{3}\right)$ and is graphed below.

$$0 \quad 1 \quad 2 \quad \overset{\frac{7}{3}}{} \quad 3 \quad 4$$

REMEMBER IT

The symbol ∞ represents *infinity*.

Your Turn Solve each inequality. Graph each solution on a number line.

a. $-3x \geq 21$

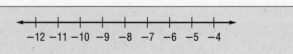

$$-12 \quad -11 \quad -10 \quad -9 \quad -8 \quad -7 \quad -6 \quad -5 \quad -4$$

b. $-2x > \dfrac{x+5}{3}$

EXAMPLE Write an Inequality

4 **CONSUMER COSTS** Alida has at most $10.50 to spend at a convenience store. She buys a bag of potato chips and a can of soda for $1.55. If gasoline at this store costs $1.35 per gallon, how many gallons of gasoline can Alida buy for her car, to the nearest tenth of a gallon?

Let g = the gallons of gasoline Alida can buy for her car. Write and solve an inequality.

$$1.55 + 1.35g \le 10.50$$ Original inequality

$1.55 + 1.35g -$ ☐ $\le 10.50 -$ ☐   Subtract from each side.

☐ $\le$ ☐   Simplify.

☐ $\le$ ☐   Divide.

$g \le$ ☐   Simplify.

Alida can buy up to ☐ gallons of gasoline for her car.

Your Turn Jeb wants to rent a car for his vacation. Value Cars rents cars for $25 per day plus $0.25 per mile. How far can he drive for one day if he wants to spend no more that $200 on car rental?

HOMEWORK ASSIGNMENT

Page(s):

Exercises:

Solving Compound and Absolute Value Inequalities

WHAT YOU'LL LEARN

- Solve compound inequalities.
- Solve absolute value inequalities.

BUILD YOUR VOCABULARY (page 2)

A **compound inequality** consists of two inequalities joined by the word [ ] or the word [ ].

The graph of a compound inequality containing *and* is the **intersection** of the solution sets of the two inequalities.

EXAMPLE Solve an "and" Compound Inequality

1 Solve $10 \leq 3y - 2 < 19$. Graph the solution set.

METHOD 1 Write the compound inequality using the word *and*. Then solve each inequality.

$$10 \leq 3y - 2 \qquad \text{and} \qquad 3y - 2 < 19$$

$$\boxed{} \leq 3y \qquad\qquad 3y < \boxed{}$$

$$\boxed{} \leq y \qquad\qquad y < \boxed{}$$

$$\boxed{} \leq y < \boxed{}$$

KEY CONCEPT

"And" Compound Inequalities A compound inequality containing the word *and* is true if and only if *both* inequalities are true.

FOLDABLES Write this concept in your notes.

METHOD 2 Solve both parts at the same time by adding 2 to each part. Then divide each part by 3.

$$\boxed{} \leq 3y - 2 < \boxed{}$$

$$\boxed{} \leq 3y < \boxed{}$$

$$\boxed{} \leq y < \boxed{}$$

Graph each solution set. Then find their intersection.

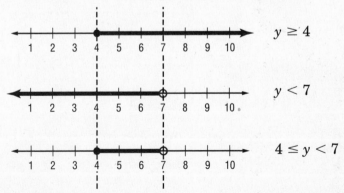

$y \geq 4$

$y < 7$

$4 \leq y < 7$

The solution set is $\{y \mid 4 \leq y < 7\}$.

Your Turn Solve $11 \le 2x + 5 < 17$. Graph the solution set.

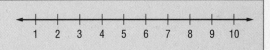

BUILD YOUR VOCABULARY (page 3)

The graph of a compound inequality containing ⬜

is the **union** of the solution sets of the two inequalities.

KEY CONCEPT

"Or" Compound Inequalities A compound inequality containing the word *or* is true if one or more of the inequalities is true.

FOLDABLES Write this concept in your notes.

EXAMPLE Solve an "or" Compound Inequality

2 Solve $x + 3 < 2$ or $-x \le -4$. Graph the solution set.

Solve each inequality separately.

$x + 3 < 2$ or $-x \le -4$

$x <$ ⬜ $x \ge$ ⬜

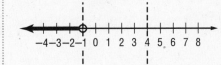

$x <$ ⬜

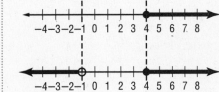

$x \ge$ ⬜

$x <$ ⬜ or $x \ge$ ⬜

The solution set is $\{x \mid x < -1 \text{ or } x \ge 4\}$.

Your Turn Solve $x + 5 < 1$ or $-2x \le -6$. Graph the solution set.

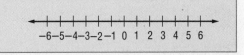

REMEMBER IT

Compound inequalities containing *and* are *conjunctions*. Compound inequalities containing *or* are *disjunctions*.

EXAMPLE Solve an Absolute Value Inequality ($<$)

3 Solve $3 > |d|$. Graph the solution set on a number line.

You can interpret $3 > |d|$ to mean that the distance between

d and 0 on a number line is less than ⬜ units. To make

$3 > |d|$ true, you must substitute numbers for d that are fewer than 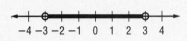 units from ☐ .

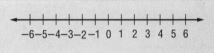

Notice that the graph of $3 > |d|$ is the same as the graph of $d >$ ☐ or $d <$ ☐ .

All of the numbers *not* at or between -3 and 3 are less than ☐ units from 0. The solution set is ☐ .

Your Turn Solve $|x| < 5$. Graph the solution.

—————————————————————
-6-5-4-3-2-1 0 1 2 3 4 5 6

EXAMPLE Solve a Multi-Step Absolute Value Inequality

4 Solve $|2x - 2| \geq 4$. Graph the solution set.

$|2x - 2| \geq 4$ is equivalent to $2x - 2 \geq 4$ or $2x - 2 \leq -4$. Solve each inequality.

$2x - 2 \geq 4$ or $2x - 2 \leq -4$

☐ $\geq$ ☐   ☐ $\leq$ ☐

$x \geq$ ☐ $x \leq$ ☐

The solution set is 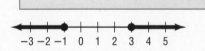 .

—————●———————●———
-3 -2 -1 0 1 2 3 4 5

Your Turn Solve $|3x - 3| > 9$. Graph the solution set.

-4 -3 -2 -1 0 1 2 3 4 5 6

BRINGING IT ALL TOGETHER

STUDY GUIDE

FOLDABLES™	**VOCABULARY PUZZLEMAKER**	**BUILD YOUR VOCABULARY**
Use your **Chapter 1 Foldable** to help you study for your chapter test.	To make a crossword puzzle, word search, or jumble puzzle of the vocabulary words in Chapter 1, go to: www.glencoe.com/sec/math/t_resources/free/index.php	You can use your completed **Vocabulary Builder** (pages 2–3) to help you solve the puzzle.

1-1

Expressions and Formulas

1. Find the value of $30 - 4^2 \div 2 \cdot 4$.

2. Evaluate $2x^2 - 3xy$ if $x = -4$ and $y = 5$.

3. Why is it important for everyone to use the same order of operations for evaluating expressions?

1-2

Properties of Real Numbers

4. Name the sets of numbers to which $-\dfrac{7}{8}$ belongs.

5. Write the Associative Property of Addition in symbols. Then illustrate this property by finding the sum $12 + 18 + 45$.

Complete each sentence.

6. The [ ] Property of Addition says that adding 0 to any number does not change its value.

7. The [ ] numbers can be written as ratios of two integers, with the integer in the denominator not being 0.

1-3
Solving Equations

Solve each equation. Check your solution.

8. $3 - 5y = 4y + 6$

9. $\dfrac{2}{3} - \dfrac{1}{2}x = \dfrac{1}{6}$

10. Solve $A = \dfrac{1}{2}h\,(a + b)$ for h.

Read the following problem and then write an equation that you could use to solve it. Do not actually solve the equation. In your equation, let *m* be the number of miles driven.

11. When Louisa rented a moving truck, she agreed to pay $28 per day plus $0.42 per mile. If she kept the truck for 3 days and the rental charges (without tax) were $153.72, how many miles did Louisa drive the truck?

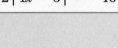

1-4
Solving Absolute Value Equations

12. Evaluate $|m - 5n|$ if $m = -3$ and $n = 2$.

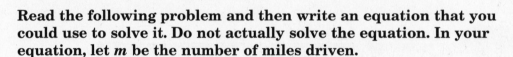

Solve each equation.

13. $-2\,|4x - 5| = -46$

14. $|7 + 3x| = x - 5$

15. Explain why the absolute value of a number can never be negative.

1-5

Solving Inequalities

There are several different ways to write or show solution sets of inequalities. Write each of the following in interval notation.

16. $\{x \mid x < -3\}$

17.

Solve each inequality. Graph the solution set.

18. $5y + 9 > 34$

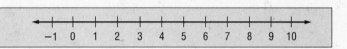

19. $-1 - 5x \le 4(x + 2)$

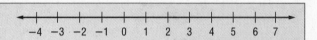

1-6

Solving Compound and Absolute Value Inequalities

Complete each sentence.

20. Two inequalities combined by the word *and* or the word *or* form a _____ .

21. The graph of a compound inequality containing the word *and* is the _____ of the graphs of the two separate inequalities.

Solve each inequality. Graph the solution set.

22. $-11 < 3m - 2 < 22$

23. $|x + 3| \ge 1$

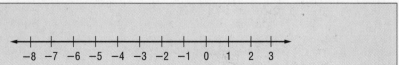

ARE YOU READY FOR THE CHAPTER TEST?

Math Online

Visit **algebra2.com** to access your textbook, more examples, self-check quizzes, and practice tests to help you study the concepts in Chapter 1.

Check the one that applies. Suggestions to help you study are given with each item.

☐ **I completed the review of all or most lessons without using my notes or asking for help.**

- You are probably ready for the Chapter Test.

- You may want to take the Chapter 1 Practice Test on page 51 of your textbook as a final check.

☐ **I used my Foldable or Study Notebook to complete the review of all or most lessons.**

- You should complete the Chapter 1 Study Guide and Review on pages 47–50 of your textbook.

- If you are unsure of any concepts or skills, refer back to the specific lesson(s).

- You may also want to take the Chapter 1 Practice Test on page 51 of your textbook.

☐ **I asked for help from someone else to complete the review of all or most lessons.**

- You should review the examples and concepts in your Study Notebook and Chapter 1 Foldable.

- Then complete the Chapter 1 Study Guide and Review on pages 47–50 of your textbook.

- If you are unsure of any concepts or skills, refer back to the specific lesson(s).

- You may also want to take the Chapter 1 Practice Test on page 51 of your textbook.

Student Signature	Parent/Guardian Signature

Teacher Signature

 **Linear Relations and Functions**

 Use the instructions below to make a Foldable to help you organize your notes as you study the chapter. You will see Foldable reminders in the margin of this Interactive Study Notebook to help you in taking notes.

Begin with two sheets of grid paper.

STEP 1 **Fold**
Fold in half along the width and staple along the fold.

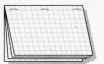

STEP 2 **Cut and Label**
Cut the top three sheets and label as shown.

 NOTE-TAKING TIP: When taking notes, make annotations. Annotations are usually notes taken in the margins of books you own to organize the text for review or study.

Chapter 2

BUILD YOUR VOCABULARY

This is an alphabetical list of new vocabulary terms you will learn in Chapter 2. As you complete the study notes for the chapter, you will see Build Your Vocabulary reminders to complete each term's definition or description on these pages. Remember to add the textbook page number in the second column for reference when you study.

Vocabulary Term	Found on Page	Definition	Description or Example
absolute value function			
boundary			
constant function			
family of graphs			
function			
greatest integer function			
identity function			
linear equation			
line of fit			
one-to-one function			

Vocabulary Term	Found on Page	Definition	Description or Example
parent graph			
piecewise function [PEES-wyz]			
point-slope form			
prediction equation [pree-DIHK-shuhn]			
relation			
scatter plot			
slope			
slope-intercept form [IHN-tuhr-SEHPT]			
standard form			
step function			

Relations and Functions

WHAT YOU'LL LEARN

- Analyze and graph relations.
- Find functional values.

BUILD YOUR VOCABULARY (pages 24–25)

A **relation** is a set of [   ].

A **function** is a special type of relation in which each

element of the domain is paired with [   ]

element of the range.

A function where each element of the [   ] is paired

with exactly one element of the [   ] is called a

one-to-one function.

EXAMPLE   Domain and Range

1 State the domain and range of the relation shown in the
graph. Is the relation a function?

The relation is {(1, 2), (3, 3), (0, −2),
(−4, 0), (−3, 1)}.

The domain is [   ].

The range is [   ].

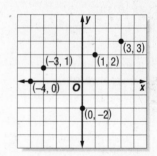

Each member of the domain is paired with exactly one
member of the range, so this relation is a function.

Your Turn   State the domain and range of the relation
shown in the graph. Is the relation a function?

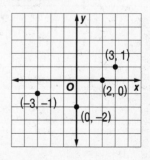

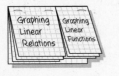

ORGANIZE IT

Under the tab for Graphing Linear Functions, define function and relation. Then draw a graph of a function and a relation.

EXAMPLE Graph Is a Line

2 **a. Graph the relation represented by $y = 3x - 1$.**

Make a table of values to find ordered pairs that satisfy the equation. Choose values for x and find the corresponding values for y. Then graph the ordered pairs.

x	y
−1	
0	
1	
2	

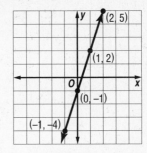

b. Find the domain and range.

The domain and range are both all [ ] numbers.

c. Determine whether the relation is a function.

This graph passes the vertical line test. For each

[ ]-value, there is exactly one [ ]-value.

The equation [ ] represents a [ ].

KEY CONCEPT

Vertical Line Test If no vertical line intersects a graph in more than one point, the graph represents a function. If some vertical line intersects a graph in two or more points, the graph does not represent a function.

Your Turn

a. Graph $y = 2x + 5$.

b. Find the domain and range.

c. Determine whether the relation is a function.

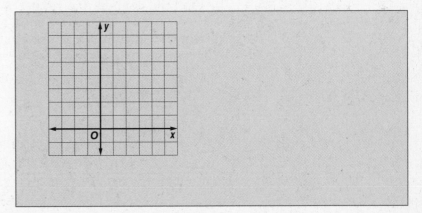

EXAMPLE Graph Is a Curve

3 **a. Graph the relation represented by $x = y^2 + 1$.**

Make a table. In this case, it is easier to choose y-values and then find the corresponding values for x. Then sketch the graph, connecting the points with a smooth curve.

x	y
	−2
	−1
	0
	1
	2

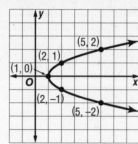

b. Find the domain and range.

The domain is $\{x \mid x \geq 1\}$. The 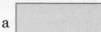 is all real numbers.

c. Determine whether the relation is a function.

You can see from the table and the vertical line test that there are two y-values for each x-value except $x = 1$.

The equation $x = y^2 + 1$ does not represent a ⬚.

Your Turn

a. Graph $x = y^2 - 3$.

b. Find the domain and range.

c. Determine whether the relation is a function.

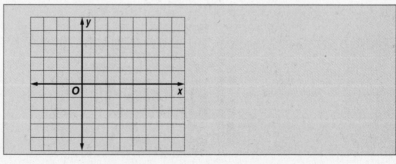

© Glencoe/McGraw-Hill

EXAMPLE Evaluate a Function

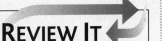

REVIEW IT

Explain what it means to evaluate an expression. *(Lesson 1-1)*

_____

_____

_____

_____

_____

_____

④ Given $f(x) = x^3 - 3$ **and** $h(x) = 0.3x^2 - 3x - 2.7$, **find each value.**

a. $f(-2)$

$f(x) = x^3 - 3$ Original function

$f(-2) = $ ⬜   Substitute.

$= $ ⬜ **or** ⬜   Simplify.

b. $h(1.6)$

$h(x) = 0.3x^2 - 3x - 2.7$ Original function

$h(1.6) = $ ⬜   Substitute.

$= $ ⬜   Multiply.

$= $ ⬜   Simplify.

c. $f(2t)$

$f(x) = x^3 - 3$ Original function

$f(2t) = $ ⬜   Substitute.

$= $ ⬜ $(ab)^2 = a^2b^2$

Your Turn Given $f(x) = x^2 + 5$ and $h(x) = 0.5x^2 + 2x + 2.5$, find each value.

a. $f(-1)$

b. $h(1.5)$

c. $f(3a)$

⬜

HOMEWORK ASSIGNMENT

Page(s): _____

Exercises: _____

Linear Equations

WHAT YOU'LL LEARN

- Identify linear equations and functions.

- Write linear equations in standard form and graph them.

BUILD YOUR VOCABULARY (page 24)

An equation such as $x + y = 4$ is called a linear equation.

A **linear equation** has no operations other than

[ ] , [ ] , and [ ]

of a variable by a constant.

EXAMPLE Identify Linear Functions

WRITE IT

Give an example of a linear function and a nonlinear function. Explain how you can tell the difference between the two functions.

_____

_____

_____

_____

_____

1 State whether each function is a linear function. Explain.

a. $g(x) = 2x - 5$

This is a [ ] function because it is in the form

$g(x) = mx + b$. $m =$ [ ] , $b =$ [ ]

b. $p(x) = x^3 + 2$

This is not a linear function because x has an exponent

other than [ ] .

Your Turn State whether each function is a linear function. Explain.

a. $h(x) = 3x - 2$

b. $g(x, y) = 3xy$

EXAMPLE Standard Form

KEY CONCEPT

Standard Form of a Linear Equation The standard form of a linear equation is $Ax + By = C$, where $A \geq 0$, A and B are not both zero.

2 Write each equation in standard form. Identify A, B, and C.

a. $y = 3x - 9$

$$y = 3x - 9$$ Original equation

 =    Subtract $3x$ from each side.

☐ = ☐   Multiply each side by -1 so that $A \geq 0$.

So, $A =$ ☐ , $B =$ ☐ , and $C =$ ☐ .

b. $-\dfrac{2}{3}x = 2y - 1$

$$-\frac{2}{3}x = 2y - 1$$ Original equation

☐ = ☐   Subtract $2y$ from each side.

☐ = ☐   Multiply each side by -3 so that the coefficients are all integers.

So, $A =$ ☐ , $B =$ ☐ , and $C =$ ☐ .

Your Turn Write each equation in standard form. Identify A, B, and C.

a. $y = -2x + 5$

b. $\dfrac{3}{5}x = -3y + 2$

c. $3x - 9y + 6 = 0$

EXAMPLE Use Intercepts to Graph a Line

③ **Find the x-intercept and the y-intercept of the graph of $-2x + y - 4 = 0$. Then graph the equation.**

The x-intercept is the value of x when $y = 0$.

$-2x + y - 4 = 0$ Original equation

$-2x + \boxed{} - 4 = 0$ Substitute 0 for y.

$\boxed{} = \boxed{}$ Add 4 to each side.

$x = \boxed{}$ Divide each side by -2.

The x-intercept is -2. The graph crosses the x-axis at $\boxed{}$.

Likewise, the y-intercept is the value of y when $x = 0$.

$-2x + y - 4 = 0$ Original equation

$-2\left(\boxed{}\right) + y - 4 = 0$ Substitute 0 for x.

$y = \boxed{}$ Add 4 to each side.

The y-intercept is $\boxed{}$. The graph crosses the y-axis at $\boxed{}$. Use the ordered pairs to graph this equation.

The x-intercept is -2, and the y-intercept is 4.

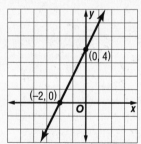

REMEMBER IT

An equation such as $x = 3$ represents a vertical line, and only has an x-intercept.

An equation such as $y = 5$ represents a horizontal line, and only has a y-intercept.

Your Turn Find the x-intercept and the y-intercept of the graph of $3x - y + 6 = 0$. Then graph the equation.

HOMEWORK ASSIGNMENT

Page(s):
Exercises:

Slope

WHAT YOU'LL LEARN

- Find and use the slope of a line.
- Graph parallel and perpendicular lines.

KEY CONCEPT

Slope of a Line The slope of a line is the ratio of the change in y-coordinates to the change in x-coordinates.

EXAMPLE Find Slope

① **Find the slope of the line that passes through (1, 3) and (−2, −3). Then graph the line.**

$$m = \frac{y_2 - y_1}{x_2 - x_1} \qquad \text{Slope formula}$$

$$= \boxed{} \qquad (x_1, y_1) = (1, 3), (x_2, y_2) = (-2, -3)$$

$$= \boxed{} \text{ or } \boxed{} \qquad \text{Simplify.}$$

Graph the two ordered pairs and draw the line. Use the slope to check your graph by selecting any $\boxed{}$ on the line. Then go up $\boxed{}$ units and right $\boxed{}$ unit or go $\boxed{}$ 2 units and left 1 unit. This point should also be on the line.

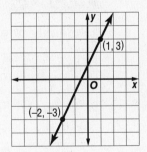

EXAMPLE Use Slope to Graph a Line

② **Graph the line passing through (1, −3) with a slope of $-\frac{3}{4}$.**

Graph the ordered pair (1, −3). Then, according to the slope, go down $\boxed{}$ units and right $\boxed{}$ units.

Plot the new point at $\boxed{}$.

Draw the line containing the points.

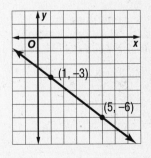

Your Turn

a. Find the slope of the line that passes through (2, 3) and (−1, 5). Then graph the line.

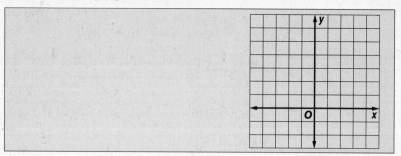

b. Graph the line passing through (2, 5) with a slope of −3.

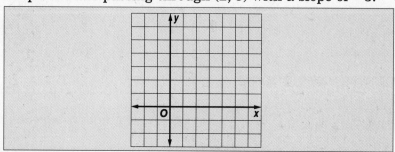

BUILD YOUR VOCABULARY (pages 24–25)

A **family of graphs** is a group of graphs that displays one or more [ ].

The **parent graph** is the [ ] of the graphs in a family.

EXAMPLE Parallel Lines

KEY CONCEPT

Parallel Lines In a plane, nonvertical lines with the same slope are parallel. All vertical lines are parallel.

3 Graph the line through (1, −2) that is parallel to the line with the equation $x - y = -2$.

The x-intercept is [ ], and the

y-intercept is [ ].

Use the intercepts to graph $x - y = -2$.

The line rises 1 unit for every 1 unit it moves to the right,

so the slope is [ ].

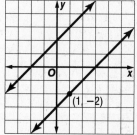

Now, use the slope and the point at (1, −2) to graph the line parallel to $x - y = -2$.

Your Turn Graph the line through (2, 3) that is parallel to the line with the equation $3x + y = 6$.

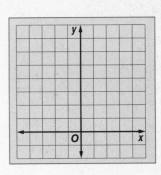

EXAMPLE Perpendicular Line

④ **Graph the line through (2, 1) that is perpendicular to the line with the equation $2x - 3y = 3$.**

The x-intercept is [ ] or [ ], and the y-intercept is

[ ]. Use the intercepts to graph $2x - 3y = 3$.

The line rises 1 unit for every 1.5 units it moves to the right,

so the slope is [ ] or [ ].

The slope of the line perpendicular

is the opposite reciprocal of [ ]

or [ ]. Start at (2, 1) and go

down [ ] units and right [ ] units.

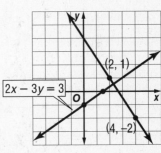

$2x - 3y = 3$

(2, 1)

(4, −2)

Use this point and (2, 1) to graph the line.

Your Turn Graph the line through (−3, 1) that is perpendicular to the line with the equation $5x - 10y = -20$.

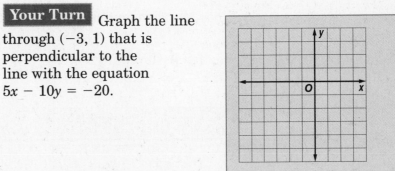

HOMEWORK ASSIGNMENT

Page(s): _____

Exercises: _____

Writing Linear Equations

Write an Equation Given Slope and a Point

WHAT YOU'LL LEARN

- Write an equation of a line given the slope and a point on the line.

- Write an equation of a line parallel or perpendicular to a given line.

KEY CONCEPTS

Slope-Intercept Form of a Linear Equation The slope-intercept form of the equation of a line is $y = mx + b$, where m is the slope and b is the y-intercept.

Point-Slope Form of a Linear Equation The point-slope form of the equation of a line is $y - y_1 = m(x - x_1)$, where (x_1, y_1) are the coordinates of a point on the line and m is the slope of the line.

① Write an equation in slope-intercept form for the line that has a slope of $-\frac{3}{5}$ and passes through $(5, -2)$.

$y = mx + b$ Slope-intercept form

$-2 = -\frac{3}{5}(5) + b$ $(x, y) = (5, 2)$, $m = -\frac{3}{5}$

$\boxed{} = \boxed{}$ Simplify.

$\boxed{} = \boxed{}$ Add 3 to each side.

The y-intercept is $\boxed{}$. So the equation in slope-intercept

form is $\boxed{}$.

Write an Equation Given Two Points

② What is an equation of the line through $(2, -3)$ and $(-3, 7)$?

First, find the slope of the line.

$m = \dfrac{y_2 - y_1}{x_2 - x_1}$ Slope formula

$= \dfrac{\boxed{} - \left(\boxed{}\right)}{\boxed{} - \boxed{}}$ $(x_1, y_1) = (2, -3)$, $(x_2, y_2) = (-3, 7)$

$= \boxed{}$ or $\boxed{}$ Simplify.

Then use the point-slope formula to find an equation.

$y - y_1 = m(x - x_1)$ Point-slope form

$y - (-3) = -2(x - 2)$ $m = -2$; you can use either point for (x_1, y_1).

$\boxed{} = \boxed{}$ Distributive Property

$y = \boxed{}$ Subtract 3 from each side.

EXAMPLE Write an Equation of a Perpendicular Line

3 **Write an equation for the line that passes through (3, −2) and is perpendicular to the line whose equation is $y = -5x + 1$.**

The slope of the given line is [   ]. Since the slopes of perpendicular lines are opposite reciprocals, the slope of the perpendicular line is [   ].

Use the point-slope form and the ordered pair [   ] to write the equation.

$$y - y_1 = m(x - x_1)$$ Point-slope form

$$y - \left(\boxed{} \right) = \boxed{}\left(x - \boxed{} \right)$$ $(x_1, y_1) = (3, -2)$, $m = \dfrac{1}{5}$

$$\boxed{} = \boxed{}$$ Distributive Property

$$y = \boxed{}$$ Subtract 2 from each side.

An equation of the line is [   ].

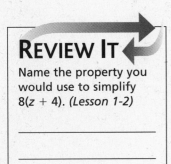

REVIEW IT

Name the property you would use to simplify $8(z + 4)$. *(Lesson 1-2)*

_____

_____

Your Turn

a. Write an equation in slope-intercept form for the line that has a slope of $\dfrac{2}{3}$ and passes through $(-3, -1)$.

[   ]

b. What is an equation of the line through $(2, 5)$ and $(-1, 3)$?

[   ]

c. Write an equation for the line that passes through $(3, 5)$ and is perpendicular to the line whose equation is $y = 3x - 2$.

[   ]

HOMEWORK ASSIGNMENT

Page(s): _____

Exercises: _____

_____

_____

Modeling Real-World Data: Using Scatter Plots

© Glencoe/McGraw-Hill

EXAMPLE Draw a Scatter Plot

WHAT YOU'LL LEARN

• Draw scatter plots.

• Find and use prediction equations.

① EDUCATION The table below shows the approximate percent of students who sent applications to two colleges in various years since 1985. Make a scatter plot of the data.

Years Since 1985	0	3	6	9	12	15
Percent	20	18	15	15	14	13

Source: *U.S. News & World Report*

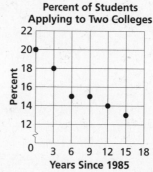

Percent of Students Applying to Two Colleges

Graph the data as ordered pairs, with the number of years since 1985 on the [   ] axis and the percentage on the [   ] axis.

Your Turn The table below shows the approximate percent of drivers who wear seat belts in various years since 1994. Make a scatter plot of the data.

Years Since 1994	0	1	2	3	4	5	6	7
Percent	57	58	61	64	69	68	71	73

Source: National Highway Traffic Safety Administration

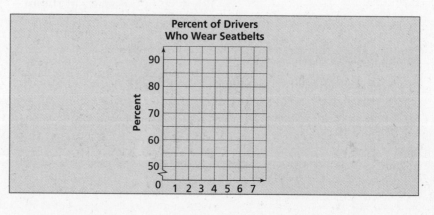

Percent of Drivers Who Wear Seatbelts

EXAMPLE Find and Use a Prediction Equation

WRITE IT

An outlier is a data point that does not appear to belong with the rest of the set. Should you include an outlier when finding a line of fit? Explain.

_____

_____

_____

_____

_____

2 EDUCATION The table and scatter plot below show the approximate percent of students who sent applications to two colleges in various years since 1985. A line of fit for the data is drawn.

Years Since 1985	0	3	6	9	12	15
Percent	20	18	15	15	14	13

Source: *U.S. News & World Report*

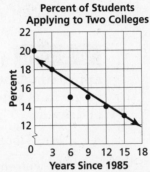

Percent of Students Applying to Two Colleges

a. Find a prediction equation. What do the slope and *y*-intercept indicate?

Find an equation of the line through (3, 18) and (15, 13). Begin by finding the slope.

$$m = \frac{y_2 - y_1}{x_2 - x_1}$$ Slope formula

$$= \frac{13 - 18}{15 - 3}$$ Substitute.

$$= \boxed{} \approx \boxed{}$$ Simplify.

$$y - y_1 = m(x - x_1)$$ Point-slope form

$$y - \boxed{} = \boxed{}\left(x - \boxed{}\right)$$ $m = -0.42$, $(x_1, y_1) = (3, 18)$

$$\boxed{} = \boxed{}$$ Distributive Property

$$y = \boxed{}$$ Add 18 to each side.

The slope indicates that the percent of students sending applications to two colleges is falling at about 0.4% each year. The *y*-intercept indicates that the percent in 1985 should have been about 19%.

b. Predict the percent in 2010.

The year 2010 is 25 years after 1985, so use the prediction equation to find the value of *y* when *x* = 25.

$y = -0.42x + 19.26$ Prediction equation

$= -0.42\left(\boxed{}\right) + 19.26$ $x = \boxed{}$

$= \boxed{}$ Simplify.

The model predicts that the percent in 2010 should be about 9%.

Your Turn The table and scatter plot show the approximate percent of drivers who wear seat belts in various years since 1994. A line of fit for the data is drawn.

Years Since 1994	0	1	2	3	4	5	6	7
Percent	57	58	61	64	69	68	71	73

Source: National Highway Traffic Safety Administration

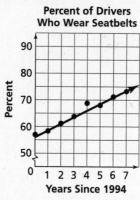

Percent of Drivers Who Wear Seatbelts

a. Find a prediction equation. What do the slope and *y*-intercept indicate?

b. Predict the percent of drivers who will be wearing seat belts in 2005.

Glencoe Algebra 2

Special Functions

WHAT YOU'LL LEARN

- Identify and graph step, constant, and identity functions.

- Identify and graph absolute value and piecewise functions.

BUILD YOUR VOCABULARY (pages 24–25)

A function whose graph is a series of line segments is called a **step function**.

The **greatest integer function**, written $f(x) = [[x]]$, is an example of a step function.

$f(x) = b$ is called a **constant function**.

When a function does not change the [ ] value, $f(x) = x$ is called the **identity function**.

Another special function is the **absolute value function**, $f(x) = |x|$.

A [ ] that is written using two or more expressions is called a **piecewise function**.

EXAMPLE Step Function

1 **PSYCHOLOGY** One psychologist charges for counseling sessions at the rate of $85 per hour or any fraction thereof. Draw a graph that represents this situation.

Use the pattern of times and costs to make a table, where x is the number of hours of the session and $C(x)$ is the total cost. Then draw the graph.

x	C(x)
$0 < x \le 1$	
$1 < x \le 2$	
$2 < x \le 3$	
$3 < x \le 4$	
$4 < x \le 5$	

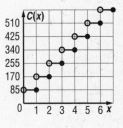

EXAMPLE Constant Function

2 **Graph $g(x) = -3$.**

For every value of x, $g(x) = -3$. The graph is a

[   ] line.

REMEMBER IT

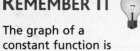

The graph of a constant function is always a horizontal line.

$g(x) = -3$	
x	$g(x)$
-2	-3
0	-3
1	-3
0.5	-3

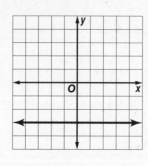

Your Turn **Graph each function.**

a. The Daily Grind charges $1.25 per pound of meat or any fraction thereof.

b. $h(x) = 2$

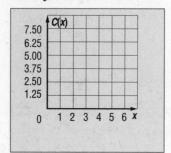

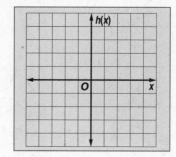

EXAMPLE Absolute Value Functions

WRITE IT

Consider the general form of an absolute value function, $f(x) = |x|$. Explain why the range of the function only includes nonnegative numbers.

_____

_____

_____

_____

3 **Graph $f(x) = |x - 3|$ and $g(x) = |x + 2|$ on the same coordinate plane. Determine the similarities and differences in the two graphs.**

Find several ordered pairs for each function.

x	$|x - 3|$
0	3
1	
2	
3	
4	

x	$|x + 2|$
-4	2
-3	
-2	
-1	
0	

Graph the points and connect them.

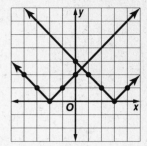

- The domain of both graphs is all ⬚ numbers.

- The range of both graphs is $\left\{y \mid y \geq \boxed{}\right\}$.

- The graphs have the same shape but different ⬚.

- The graph of $g(x)$ is the graph of $f(x)$ translated left ⬚ units.

Your Turn

a. Graph $f(x) = |x| - 2$ and $g(x) = |x| + 3$ on the same coordinate plane. Determine the similarities and differences in the two graphs.

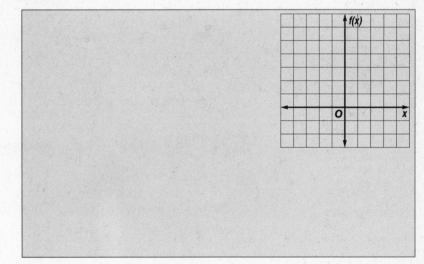

b. Graph the piecewise function $f(x) = \begin{cases} 2x + 1 & \text{if } x > -1 \\ -3 & \text{if } x \leq -1 \end{cases}$.

Identify the domain and range.

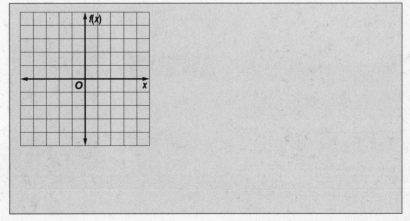

FOLDABLES™

ORGANIZE IT

Under the tab for Graphing Linear Functions, graph an example of each of the following functions: step, constant, absolute value, and piecewise.

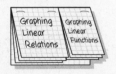

HOMEWORK ASSIGNMENT

Page(s): _____
Exercises: _____

Graphing Inequalities

BUILD YOUR VOCABULARY (page 24)

When graphing inequalities, the graph of the line is the **boundary** of each region.

EXAMPLE Dashed Boundary

1 Graph $x - 2y < 4$.

The boundary is the graph of $x - 2y = 4$. Since the inequality symbol is $>$, the boundary will be dashed. Use the slope-intercept form, $y = \frac{1}{2}x - 2$. Test $(0, 0)$.

$x - 2y < 4$		Original inequality
$\boxed{} \overset{?}{<} \boxed{}$		$(x, y) = (0, 0)$
$0 < 4$		true

Shade the region that contains $(0, 0)$.

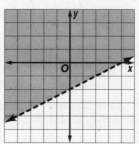

EXAMPLE Absolute Value Inequality

2 Graph $y \quad |x| - 2$.

Since the inequality symbol is $\geq$, the graph of the related equation $y = |x| - 2$ is solid. Graph the equation. Test $(0, 0)$.

$y \geq |x| - 2$

$0 \overset{?}{\geq} |0| - 2$

$0 \geq 0 - 2$ or $\boxed{} \geq \boxed{}$

Shade the region that contains $\boxed{}$.

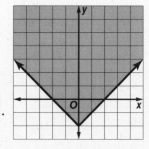

Your Turn Graph each inequality.

a. $2x + 5y > 10$

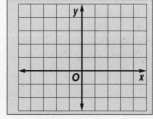

b. $y < |x| + 3$

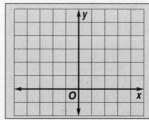

FOLDABLES™	VOCABULARY PUZZLEMAKER	BUILD YOUR VOCABULARY
Use your **Chapter 2 Foldable** to help you study for your chapter test.	To make a crossword puzzle, word search, or jumble puzzle of the vocabulary words in Chapter 2, go to: www.glencoe.com/sec/math/ t_resources/free/index.php	You can use your completed **Vocabulary Builder** (pages 24–25) to help you solve the puzzle.

2-1

Relations and Functions

For Exercises 1 and 2, refer to the graph shown at the right.

1. Write the domain and range of the relation.

D:

R:

2. Is this relation a function? Explain.

2-2

Linear Equations

3. Write $x - 2 = \frac{1}{6}y$ in standard form. Identify A, B, and C.

Write *yes* or *no* to tell whether each linear equation is in standard form. If it is not, explain why it is not.

4. $-x + 2y = 5$ **5.** $5x - 7y = 3$

2-3
Slope

6. How are the slopes of two nonvertical parallel lines related?

7. How are the slopes of two oblique perpendicular lines related?

Find the slope of the line that passes through each pair of points.

8. $(3, 7), (8, -1)$

9. $\left(8, -\dfrac{1}{4}\right), \left(0, -\dfrac{1}{4}\right)$

2-4
Writing Linear Equations

Write the equation in slope-intercept form for the line that satisfies each set of conditions.

10. slope 4, passes through $(0, 3)$

11. passes through $(5, -6)$ and $(3, 2)$

12. Write an equation for the line that passes through $(8, -5)$ and is perpendicular to the line whose equation is $y = \dfrac{1}{2}x - 8$.

2-5
Modeling Real-World Data: Using Scatter Plots

13. Draw a scatter plot for the data. Then state which of the data points is an outlier.

x	2	6	10	14	20	24
y	15	20	30	16	40	50

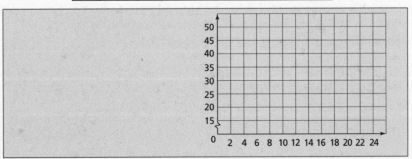

2-6

Special Functions

Write the letter of the term that best describes each function.

14. $f(x) = |4x + 3|$ ▢

15. $f(x) = [[x]] + 1$ ▢

16. $f(x) = 6$ ▢

17. $f(x) = \begin{cases} x + 3 \text{ if } x < 0 \\ 2 - x \text{ if } x \geq 0 \end{cases}$ ▢

a.	constant function
b.	absolute value function
c.	piecewise function
d.	step function

2-7

Graphing Inequalities

18. When graphing a linear inequality in two variables, how do you know whether to make the boundary a solid line or a dashed line?

19. Graph the inequality $10 - 5y < 2x$.

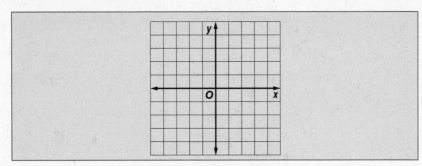

ARE YOU READY FOR THE CHAPTER TEST?

Math
Online

Visit **algebra2.com** to access your textbook, more examples, self-check quizzes, and practice tests to help you study the concepts in Chapter 2.

Check the one that applies. Suggestions to help you study are given with each item.

☐ **I completed the review of all or most lessons without using my notes or asking for help.**

• You are probably ready for the Chapter Test.

• You may want to take the Chapter 2 Practice Test on page 105 of your textbook as a final check.

☐ **I used my Foldable or Study Notebook to complete the review of all or most lessons.**

• You should complete the Chapter 2 Study Guide and Review on pages 100–104 of your textbook.

• If you are unsure of any concepts or skills, refer back to the specific lesson(s).

• You may also want to take the Chapter 2 Practice Test on page 105 of your textbook.

☐ **I asked for help from someone else to complete the review of all or most lessons.**

• You should review the examples and concepts in your Study Notebook and Chapter 2 Foldable.

• Then complete the Chapter 2 Study Guide and Review on pages 100–104 of your textbook.

• If you are unsure of any concepts or skills, refer back to the specific lesson(s).

• You may also want to take the Chapter 2 Practice Test on page 105 of your textbook.

Student Signature   Parent/Guardian Signature

Teacher Signature

Systems of Equations and Inequalities

 Use the instructions below to make a Foldable to help you organize your notes as you study the chapter. You will see Foldable reminders in the margin of this Interactive Study Notebook to help you in taking notes.

Begin with one sheet of 11" × 17" paper and four sheets of grid paper.

STEP 1 **Fold and Cut**
Fold the short sides of the 11" × 17" paper to meet in the middle. Cut each tab in half as shown.

STEP 2 **Staple and Label**
Insert 2 folded half sheets of grid paper in each tab. Staple at edges. Label each tab as shown.

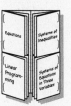

 NOTE-TAKING TIP: When taking notes, summarize the main ideas presented in the lesson. Summaries are useful for condensing data and realizing what is important.

Chapter 3

This is an alphabetical list of new vocabulary terms you will learn in Chapter 3. As you complete the study notes for the chapter, you will see Build Your Vocabulary reminders to complete each term's definition or description on these pages. Remember to add the textbook page number in the second column for reference when you study.

Vocabulary Term	Found on Page	Definition	Description or Example
bounded region			
consistent system			
constraints [kuhn-STRAYNTS]			
dependent system			
elimination method			
feasible region [FEE-zuh-buhl]			
inconsistent system [ihn-kuhn-SIHS-tuhnt]			

Vocabulary Term	Found on Page	Definition	Description or Example
independent system			
linear programming			
ordered triple			
substitution method			
system of equations			
system of inequalities			
unbounded region			
vertices			

Solving Systems of Equations by Graphing

WHAT YOU'LL LEARN

- Solve systems of linear equations by graphing.
- Determine whether a system of linear equations is consistent and independent, consistent and dependent, or inconsistent.

BUILD YOUR VOCABULARY (pages 50–51)

A **system of equations** is [   ] or more equations with the same variables.

A system of equations is **consistent** if it has at least [   ] solution and **inconsistent** if it has [   ] solutions.

A consistent system is **independent** if it has exactly [   ] solution or **dependent** if it has an [   ] number of solutions.

EXAMPLE   Solve by Graphing

1 **Solve the system of equations by graphing.**

$$x - 2y = 0$$
$$x + y = 6$$

Write each equation in slope-intercept form.

$x - 2y = 0$ → [   ]

$x + y = 6$ → [   ]

The graphs appear to intersect at [   ].

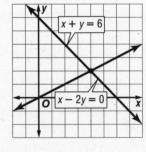

REMEMBER IT

When solving a system of equations by graphing, you should always check the ordered pair in *each* of the original equations.

Check: Substitute the coordinates into each equation.

$x - 2y = 0$	$x + y = 6$	Original equations.
$4 - 2(2) \stackrel{?}{=}$ [ ]	$4 + 2 \stackrel{?}{=} 6$	Replace *x* and *y*.
[ ] $= 0$	[ ] $= 6$	Simplify.

So, the solution of the system is [   ].

Your Turn Solve $x + 3y = 7$ and $x - y = 3$ by graphing.

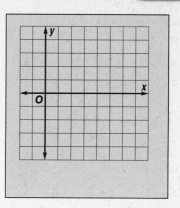

EXAMPLE Intersecting Lines

REVIEW IT
Explain the steps you would use to write $2x + 5y = 10$ in slope-intercept form. *(Lesson 1-3)*

_____

_____

_____

② **Graph the system of equations and describe it as _consistent and independent, consistent and dependent,_ or _inconsistent_.**

$x - y = 5$
$x + 2y = -4$

Write each equation in slope-intercept form.

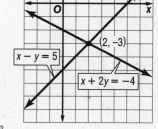

$x - y = 5 \longrightarrow$

$x + 2y = -4 \longrightarrow$

The graphs intersect at [   ]. Since there is one solution, this system is [   ] and [   ].

EXAMPLE Same Line

③ **Graph the system of equations and describe it as _consistent and independent, consistent and dependent,_ or _inconsistent_.**

$9x - 6y = -6$
$6x - 4y = -4$

$9x - 6y = -6 \longrightarrow$

$6x - 4y = -4 \longrightarrow$

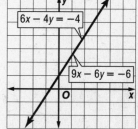

Since the equations are equivalent, their graphs are the [   ] line. There are [   ] many solutions.

This system is [   ] and [   ].

ORGANIZE IT

Under the Equations tab, graph the three possible relationships between a system of equations and the number of solutions. Write the number of solutions below each graph.

Your Turn Graph each system of equations and describe it as *consistent and independent, consistent and dependent,* or *inconsistent.*

a. $x + y = 5$
$2x = y - 11$

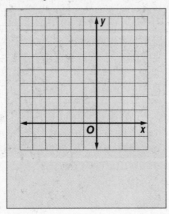

b. $x + y = 3$
$2x = -2y + 6$

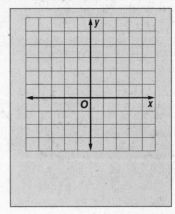

EXAMPLE Parallel Lines

4️⃣ **Graph the system of equations and describe it as *consistent and independent, consistent and dependent,* or *inconsistent.***

$15x - 6y = 0$
$5x - 2y = 10$

$15x - 6y = 0 \longrightarrow$ ⬜

$5x - 2y = 10 \longrightarrow$ ⬜

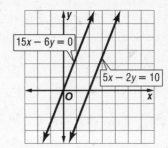

The lines do not intersect. Their graphs are parallel lines. So, there are no solutions that satisfy both equations. This system

is ⬜ .

Your Turn Graph the system of equations $y = 3x + 2$ and $-6x + 2y = 10$ and describe it as *consistent and independent, consistent and dependent,* or *inconsistent.*

HOMEWORK ASSIGNMENT

Page(s):

Exercises:

Solving Systems of Equations Algebraically

WHAT YOU'LL LEARN

- Solve systems of linear equations by using substitution.

- Solve systems of linear equations by using elimination.

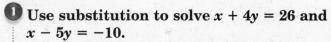

BUILD YOUR VOCABULARY (pages 50–51)

Using the **substitution method**, one equation is solved for one [ ] in terms of [ ]. Then, this expression is substituted for the variable in the other equation. Using the **elimination method**, you eliminate one of the variables by [ ] or [ ] the equations.

EXAMPLE Solve by Using Substitution

1 Use substitution to solve $x + 4y = 26$ and $x - 5y = -10$.

Solve the first equation for x in terms of y.

$x + 4y = 26$	First equation
$x = $ [ ]	Subtract $4y$ from each side.

Substitute $26 - 4y$ for x in the second equation and solve for y.

$x - 5y = -10$	Second equation
[ ] $- 5y = -10$	Substitute for x.
[ ] $=$ [ ]	Subtract from each side.
$y = $ [ ]	Divide each side.

REMEMBER IT

In Example 1, you can substitute the value for y in either of the original equations. Choose the equation that is easiest to solve.

Now substitute the value for y in either of the original equations and solve for x.

$x + 4y = 26$	First equation
$x + 4$ [ ] $= 26$	Replace y with 4.
$x + $ [ ] $= 26$	Simplify.
$x = $ [ ]	Subtract from each side.

The solution of the system is [ ].

EXAMPLE Solve by Using Elimination

2 Use the elimination method to solve $x + 2y = 10$ and $x + y = 6$.

In each equation, the coefficient of x is 1. If one equation is subtracted from the other, the variable x will be eliminated.

$$x + 2y = 10$$
$$(-)x + y = 6$$

$$y = \boxed{} \qquad \text{Subtract the equations.}$$

Now find x by substituting $\boxed{}$ for y in either original equation.

$$x + y = 6 \qquad \text{Second equation}$$

$$x + \boxed{} = 6 \qquad \text{Replace } y \text{ with } \boxed{}.$$

$$x = \boxed{} \qquad \text{Subtract } \boxed{} \text{ from each side.}$$

The solution is $\boxed{}$.

EXAMPLE Multiply, Then Use Elimination

3 Use the elimination method to solve $2x + 3y = 12$ and $5x - 2y = 11$.

Multiply the first equation by 2 and the second equation by 3. Then add the equations to eliminate the variable.

$$2x + 3y = 12 \xrightarrow{\text{Multiply by 2.}} \boxed{} + \boxed{} = \boxed{}$$

$$5x - 2y = 11 \xrightarrow{\text{Multiply by 3.}} (+)\boxed{} - \boxed{} = \boxed{}$$

$$\phantom{5x - 2y = 11 \xrightarrow{\text{Multiply by 3.}} } 19x \phantom{- \boxed{00}} = 57$$

$$\phantom{5x - 2y = 11 \xrightarrow{\text{Multiply by 3.}} 19x } x = 3$$

Replace x and solve for y.

$$2x + 3y = 12 \qquad \text{First equation}$$

$$2\left(\boxed{}\right) + 3y = 12 \qquad \text{Replace } x \text{ with } \boxed{}.$$

$$\boxed{} + 3y = 12 \qquad \text{Multiply.}$$

$$3y = 6 \qquad \text{Subtract 6 from each side.}$$

$$y = 2 \qquad \text{Divide each side by 3.}$$

The solution is .

WRITE IT

When solving a system of equations, how do you choose whether to use the substitution method or the elimination method?

_____

_____

_____

_____

_____

_____

FOLDABLES

ORGANIZE IT

Under the Equations tab, write how you recognize an inconsistent system of equations. Then write how you recognize a consistent system of equations.

EXAMPLE Inconsistent System

4 Use the elimination method to solve $-3x + 5y = 12$ and $6x - 10y = -21$.

Use multiplication to eliminate x.

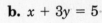

$-3x + 5y = 12$ $\xrightarrow{\text{Multiply by 2.}}$ ☐ + ☐ = ☐

$6x - 10y =$ ☐ $\qquad$ (+) ☐ − ☐ = ☐

$0 =$ ☐

Since there are no values of x and y that will make the equation

☐ = ☐ true, there are no solutions for the system

of equations.

Your Turn Use the elimination method to solve each system of equations.

a. $x - 3y = 2$
$\qquad x + 7y = 12$

b. $x + 3y = 5$
$\qquad x + 5y = -3$

c. $x + 3y = 7$
$\qquad 2x + 5y = 10$

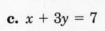

d. $2x + 3y = 11$
$\qquad -4x - 6y = 20$

HOMEWORK ASSIGNMENT

Page(s):

Exercises:

Solving Systems of Inequalities by Graphing

WHAT YOU'LL LEARN

- Solve systems of inequalities by graphing.

- Determine the coordinates of the vertices of a region formed by the graph of a system of inequalities.

FOLDABLES

ORGANIZE IT

Under the tab for Systems of Inequalities, solve the following by graphing.

$y \leq 3x - 6$

$y > -4x + 2$

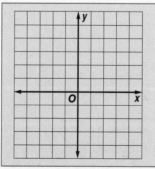

BUILD YOUR VOCABULARY (page 51)

To solve a **system of inequalities**, find the [   ] pairs that satisfy [   ] of the inequalities in the system.

EXAMPLE Intersecting Regions

1 Solve the system of inequalities $y \geq 2x - 3$ and $y < -x + 2$ by graphing.

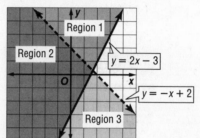

solution of $y \geq 2x - 3$ ⟶ Regions [   ]

solution of $y < -x + 2$ ⟶ Regions [   ]

The intersection of these regions is Region [   ], which is the solution of the system of inequalities. Notice that the solution is a region containing an [   ] number of ordered pairs.

Your Turn Solve each system of inequalities by graphing.

a. $y \leq 3x - 3$
     $y > x + 1$

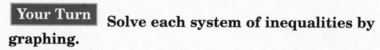

b. $y \geq -2x - 3$
     $|x + 2| < 1$

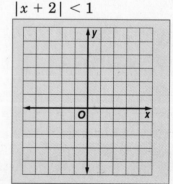

EXAMPLE Separate Regions

REMEMBER IT

You can indicate that a system of inequalities has no solutions in the following ways: empty set, null set, ∅, or {}.

2 **Solve the system of inequalities by graphing.**

$$y \geq -\frac{3}{4}x + 1$$

$$y \leq -\frac{3}{4}x - 2$$

Graph both inequalities.

The graphs do not overlap, so the solutions have no points in common.

The solution set is [ ] .

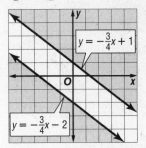

Your Turn Solve the system of inequalities $y < \frac{1}{2}x + 2$ and $y > \frac{1}{2}x + 4$ by graphing.

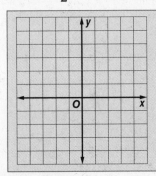

EXAMPLE Find Vertices

3 **Find the coordinates of the vertices of the figure formed by $2x - y \geq -1$, $x + y \leq 4$, and $x + 4y \geq 4$.**

Graph each inequality. The

[ ] of the graphs

forms a triangular region.

The vertices of the triangle are at

[ ] , [ ] , and [ ] .

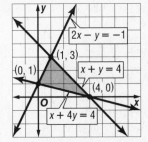

Your Turn Find the coordinates of the vertices of the figure formed by $x + 2y \geq 1$, $x + y \leq 3$, and $-2x + y \leq 3$.

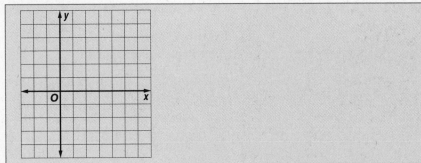

HOMEWORK ASSIGNMENT

Page(s):

Exercises:

WHAT YOU'LL LEARN

- Find the maximum and minimum values of a function over a region.

- Solve real-world problems using linear programming.

BUILD YOUR VOCABULARY (pages 50–51)

In a graph of a system of inequalities, the [____] are called the **constraints**.

The intersection of the graphs is called the **feasible region**.

When the graph of a system of constraints is a polygonal region, we say that the region is **bounded**.

The maximum or minimum value of a related function

[____] occurs at one of the **vertices** of the

feasible region.

When a system of inequalities forms a region that is

[____] , the region is said to be **unbounded**.

The process of finding [____] or [____]

values of a function for a region defined by inequalities is called **linear programming**.

FOLDABLES

ORGANIZE IT

Under the Linear Programming tab, sketch a graph of a system of inequalities like the one shown in Example 1. Then label the constraints, feasible region, and vertices of the graph.

EXAMPLE Bounded Region

1 Graph the following system of inequalities. Name the coordinates of the vertices of the feasible region. Find the maximum and minimum values of the function $f(x, y) = 3x - 2y$ for this region.

$x \leq 5$

$y \leq 4$

$x + y \geq 2$

First, find the vertices of the region. Graph the inequalities.

The polygon formed is a triangle with vertices at

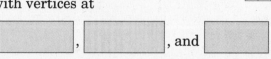

[____] , [____] , and [____] .

Next, use a table to find the maximum and minimum values of $f(x, y)$. Substitute the coordinates of the vertices into the function.

(x, y)	$3x - 2y$	$f(x, y)$	
$(-2, 4)$			← minimum
$(5, -3)$			← maximum
$(5, 4)$			

The vertices of the feasible region are [ ] , [ ] ,

and [ ] . The maximum value is [ ] at [ ] .

The minimum value is [ ] at [ ] .

WRITE IT

Explain how you recognize the unbounded region of a system of inequalities.

_____

_____

_____

_____

_____

Your Turn Graph each system of inequalities. Name the coordinates of the vertices of the feasible region. Find the maximum and minimum values of the given function for this region.

a. $x \le 4$
  $y \le 5$
  $x + y \ge 6$
  $f(x, y) = 4x - 3y$

b. $x + 3y \le 6$
  $-x - 3y \le 9$
  $2y - x \ge -6$
  $f(x, y) = x + 2y$

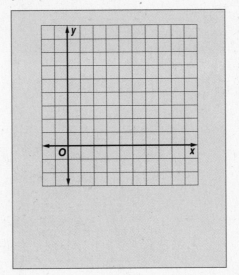

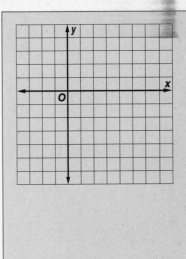

KEY CONCEPT

Linear Programming Procedure

Step 1 Define the variables.

Step 2 Write a system of inequalities.

Step 3 Graph the system of inequalities.

Step 4 Find the coordinates of the vertices of the feasible region.

Step 5 Write a function to be maximized or minimized.

Step 6 Substitute the coordinates of the vertices into the function.

Step 7 Select the greatest or least result. Answer the problem.

HOMEWORK ASSIGNMENT

Page(s):

Exercises:

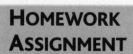

EXAMPLE **Linear Programming**

2 **LANDSCAPING** A landscaping company has crews who mow lawns and prune shrubbery. The company schedules 1 hour for mowing jobs and 3 hours for pruning jobs. Each crew is scheduled for no more than 2 pruning jobs per day. Each crew's schedule is set up for a maximum of 9 hours per day. On the average, the charge for mowing a lawn is $40 and the charge for pruning shrubbery is $120. Find a combination of mowing lawns and pruning shrubs that will maximize the income the company receives per day from one of its crews.

STEP 1 Define the variables.

m = the number of mowing jobs

p = the number of pruning jobs

STEP 2 Write a system of inequalities.

Since the number of jobs cannot be negative, m and p must be nonnegative numbers.

$m \geq$ [ ] , $p \geq$ [ ]

Mowing jobs take [ ] hour. Pruning jobs take [ ] hours. There are [ ] hours to do the jobs.

[ ] + [ ] ≤ 9

There are no more than 2 pruning jobs a day. [ ] ≤ 2

STEP 3 Graph the system of inequalities.

STEP 4 Find the coordinates of the vertices of the feasible region.

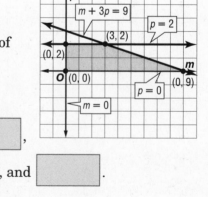

From the graph, the vertices are at [ ] , [ ] , [ ] , and [ ] .

STEP 5 Write the function to be maximized.

The function that describes the income is $f(m, p) = 40m + 120p$. We want to find the [ ] value for this function.

STEP 6 Substitute the coordinates of the vertices into the function.

(m, p)	40m + 120p	f(m, p)
(0, 2)		
(3, 2)		
(9, 0)		
(0, 0)		

STEP 7 Select the [   ] amount. Answer the problem.

The maximum values are [   ] at [   ] and [   ] at [   ]. This means that the company receives the most money with [   ] mows and [   ] prunings or [   ] mows and [   ] prunings.

Your Turn A landscaping company has crews who rake leaves and mulch. The company schedules 2 hours for mulching jobs and 4 hours for raking jobs. Each crew is scheduled for no more than 2 raking jobs per day. Each crew's schedule is set up for a maximum of 8 hours per day. On the average, the charge for raking a lawn is $50 and the charge for mulching is $30. Find a combination of raking leaves and mulching that will maximize the income the company receives per day from one of its crews.

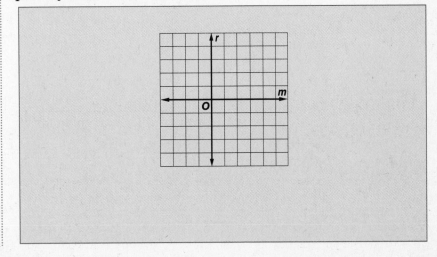

HOMEWORK ASSIGNMENT

Page(s):

Exercises:

Solving Systems of Equations in Three Variables

WHAT YOU'LL LEARN

- Solve systems of linear equations in three variables.

- Solve real-world problems using systems of linear equations in three variables.

KEY CONCEPT

System of Equations in Three Variables

One solution
- planes intersect in one point

Infinite Solutions
- planes intersect in a line
- planes intersect in the same plane

No solution
- planes have no point in common

BUILD YOUR VOCABULARY (page 51)

The solution of a system of equations in three variables, x, y, and z is called an **ordered triple** and is written as (x, y, z).

EXAMPLE One Solution

1 **Solve the system of equations.**

$5x + 3y + 2z = 2$

$2x + y - z = 5$

$x + 4y + 2z = 16$

Use elimination to make a system of two equations in two variables. First, eliminate z in the first and second equations.

$5x + 3y + 2z = 2$

$2x + y - z = 5$ $\xrightarrow{(\times 2)}$

$5x + 3y + 2z = 2$

$(+)\ 4x + 2y - 2z = 10$

$\boxed{} = \boxed{}$

Eliminate z in the first and third equations.

$5x + 3y + 2z = 2$

$(-)x + 4y + 2z = 16$

$\boxed{} = \boxed{}$

Solve the system of two equations. Eliminate y.

$9x + 5y = 12$

$4x - y = -14$ $\xrightarrow{(\times 5)}$

$9x + 5y = 12$

$(+)\ \boxed{} = \boxed{}$

$\boxed{} = \boxed{}$

$x = \boxed{}$

Substitute $\boxed{}$ for x in one of the two equations with two variables and solve for y.

$4x - y = -14$ Equation with two variables

$4\left(\boxed{}\right) - y = -14$ Replace x.

$\boxed{} - y = -14$ Multiply

$y = \boxed{}$ Simplify.

FOLDABLES

ORGANIZE IT

Under the Systems of Equations in Three Variables tab, sketch a graph of a system with
(a) one solution,
(b) infinite soultions,
(c) no solutions.

Substitute [ ] for x and y in one of the original equations with three variables.

$2x + y - z = 5$ Equation with three variables

$2\left(\boxed{}\right) + 6 - z = 5$ Replace x and y.

$\boxed{} + 6 - z = 5$ Multiply.

$z = \boxed{}$ Simplify.

The solution is [ ]. You can check this solution in the other two original equations.

EXAMPLE   No Solution

2 **Solve the system of equations.**

$3x - y - 2z = 4$

$6x + 4y + 8z = 11$

$9x + 6y + 12z = -3$

Eliminate x in the second two equations.

$6x + 4y + 8z = 11 \xrightarrow{(\times 3)} \boxed{} + 12y + \boxed{} = \boxed{}$

$9x + 6y + 12z = -3 \xrightarrow{(\times 2)} (-)18x + \boxed{} + 24z = \boxed{}$

$\boxed{} = \boxed{}$

The equation $\boxed{} = \boxed{}$ is never true. So, there is no solution of this system.

Your Turn   Solve each system of equations.

a. $2x + 3y - 3z = 16$

$x + y + z = -3$

$x - 2y - z = -1$

b. $x + y - 2z = 3$

$-3x - 3y + 6z = -9$

$2x + y - z = 6$

HOMEWORK ASSIGNMENT

Page(s):

Exercises:

BRINGING IT ALL TOGETHER

FOLDABLES™	VOCABULARY PUZZLEMAKER	BUILD YOUR VOCABULARY
Use your **Chapter 3 Foldable** to help you study for your chapter test.	To make a crossword puzzle, word search, or jumble puzzle of the vocabulary words in Chapter 3, go to: www.glencoe.com/sec/math/t_resources/free/index.php	You can use your completed **Vocabulary Builder** (pages 50–51) to help you solve the puzzle.

3-1

Solving Systems of Equations by Graphing

Under each system graphed below, write all of the following words that apply: *consistent*, *inconsistent*, *dependent*, and *independent*.

1.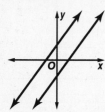

2.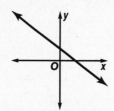

3.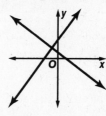

4. Solve the system $x + y = 3$ and $3x - y = 1$ by graphing.

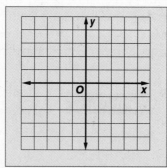

3-2

Solving Systems of Equations Algebraically

5. Solve $x = 4y + 3$ and $x - 2y = 9$ by using substitution.

6. Solve $-2x + 5y = 7$ and $2x + 4y = 11$ by using elimination.

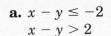

Solving Systems of Inequalities by Graphing

7. Which system of inequalities matches the graph shown at the right? ☐

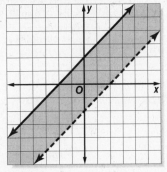

a. $x - y \le -2$
   $x - y > 2$

b. $x - y \ge -2$
   $x - y < 2$

c. $x + y \le -2$
   $x + y > 2$

d. $x - y > -2$
   $x - y \le 2$

Find the coordinates of the vertices of the figure formed by each system of inequalities.

8. $x + y \ge 8$
   $y \ge 5$
   $x \ge 0$

9. $x + y \ge 6$
   $x \le 8$
   $y \le 5$

3-4

Linear Programming

10. A polygonal feasible region has vertices at $(4, -1)$, $(3, 0)$, $(1, -6)$, and $(-2, 2)$. Find the maximum and minimum of the function $f(x, y) = -2x + y$ over this region.

3-5

Solving Systems of Equations in Three Variables

Solve each system of equations.

11. $5x + y + 4z = 9$
   $-5x + 3y + z = -15$
   $15x + 5y + 7z = 9$

12. $2x + y - 4z = -3$
   $-6x - 3y + 12z = 7$
   $x - 9y + 4z = 1$

13. The sum of three numbers is 22. The sum of the first and second numbers is 19, and the first number is 5 times the third number. Find the numbers.

ARE YOU READY FOR THE CHAPTER TEST?

Math Online

Visit **algebra2.com** to access your textbook, more examples, self-check quizzes, and practice tests to help you study the concepts in Chapter 3.

Check the one that applies. Suggestions to help you study are given with each item.

☐ **I completed the review of all or most lessons without using my notes or asking for help.**

• You are probably ready for the Chapter Test.

• You may want to take the Chapter 3 Practice Test on page 149 of your textbook as a final check.

☐ **I used my Foldable or Study Notebook to complete the review of all or most lessons.**

• You should complete the Chapter 3 Study Guide and Review on pages 145–148 of your textbook.

• If you are unsure of any concepts or skills, refer back to the specific lesson(s).

• You may also want to take the Chapter 3 Practice Test on page 149 of your textbook.

☐ **I asked for help from someone else to complete the review of all or most lessons.**

• You should review the examples and concepts in your Study Notebook and Chapter 3 Foldable.

• Then complete the Chapter 3 Study Guide and Review on pages 145–148 of your textbook.

• If you are unsure of any concepts or skills, refer back to the specific lesson(s).

• You may also want to take the Chapter 3 Practice Test on page 149 of your textbook.

Student Signature	Parent/Guardian Signature

Teacher Signature

Matrices

 Use the instructions below to make a Foldable to help you organize your notes as you study the chapter. You will see Foldable reminders in the margin of this Interactive Study Notebook to help you in taking notes.

Begin with one sheet of notebook paper.

STEP 1 Fold and Cut
Fold lengthwise to the holes.
Cut eight tabs in the top sheet.

STEP 2 Label
Label each tab with a lesson number and title.

NOTE-TAKING TIP: When you take notes, write descriptive paragraphs about your learning experiences.

BUILD YOUR VOCABULARY

This is an alphabetical list of new vocabulary terms you will learn in Chapter 4. As you complete the study notes for the chapter, you will see Build Your Vocabulary reminders to complete each term's definition or description on these pages. Remember to add the textbook page number in the second column for reference when you study.

Vocabulary Term	Found on Page	Definition	Description or Example
Cramer's Rule [KRAY-muhrs]			
determinant			
dilation [dy-LAY-shuhn]			
element			
expansion by minors			
identity matrix			
image			
inverse			
isometry [eye-SAH-muh-tree]			
matrix [MAY-trihks]			

Vocabulary Term	Found on Page	Definition	Description or Example
matrix equation			
preimage			
reflection			
rotation			
scalar multiplication [SKAY-luhr]			
square matrix			
transformation			
translation			
vertex matrix			
zero matrix			

Introduction to Matrices

WHAT YOU'LL LEARN

- Organize data in matrices.

- Solve equations involving matrices.

BUILD YOUR VOCABULARY (pages 70–71)

A **matrix** is a rectangular array of variables or constants in

[   ] and vertical columns, usually

enclosed in brackets.

Each [   ] in the matrix is called an **element**.

A matrix that has the same number of rows and

[   ] is called a **square matrix**.

Another special type of matrix is the **zero matrix**, in which

every element is [   ].

EXAMPLE Organize Data in a Matrix

REMEMBER IT

An element of a matrix can be represented by the notation a_{ij}. This refers to the element in row i, column j.

1 **COLLEGE** Kaitlin wants to attend one of three Iowa universities next year. She has gathered information about tuition (T), room and board (R/B), and enrollment (E) for the universities. Use a matrix to organize the information. Which university's total cost is lowest?

Iowa State University:

T - $3132   R/B - $4432   E - 26,845

University of Iowa:

T - $3204   R/B - $4597   E - 28,311

University of Northern Iowa:

T - $3130   R/B - $4149   E - 14,106

Organize the data into labeled columns and rows.

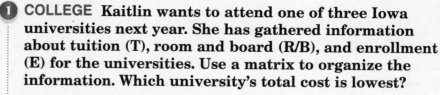

	T	R/B	E
ISU			
UI			
UNI			

The University of [   ] has the lowest total cost.

Your Turn Justin is going out for lunch. The information he has gathered from the two fast-food restaurants is listed below. Use a matrix to organize the information. When is each restaurant's total cost less expensive?

Burger Complex		Lunch Express	
Hamburger Meal	$3.39	Hamburger Meal	$3.49
Cheeseburger Meal	$3.59	Cheeseburger Meal	$3.79
Chicken Sandwich Meal	$4.99	Chicken Sandwich Meal	$4.89

FOLDABLES

ORGANIZE IT

Under the tab for Lesson 4-1, tell how to find the dimensions of a matrix.

4-1 Introduction
4-2 Operations
4-3 Multiplying Matrices
4-4 Transformations
4-5 Determinants
4-6 Cramer's Rule
4-7 Identity
4-8 Using Matrices

EXAMPLE Dimensions of a Matrix

2 **State the dimensions of matrix G if**

$$G = \begin{bmatrix} 2 & -1 & 0 & 3 \\ 1 & 5 & -3 & -1 \end{bmatrix}.$$

$$G = \begin{bmatrix} 2 & -1 & 0 & 3 \\ 1 & 5 & -3 & -1 \end{bmatrix} \Big\} \quad \boxed{} \text{ rows}$$

$$\boxed{} \text{ columns}$$

Since matrix G has $\boxed{}$ rows and $\boxed{}$ columns, the dimensions of matrix G are $\boxed{}$.

Your Turn State the dimensions of matrix G if

$$G = \begin{bmatrix} 2 & 3 \\ 0 & 4 \\ -1 & 4 \end{bmatrix}.$$

EXAMPLE Solve an Equation Involving Matrices

3 Solve $\begin{bmatrix} y \\ 3 \end{bmatrix} = \begin{bmatrix} 3x - 2 \\ 2y + x \end{bmatrix}$ for x and y.

Since the matrices are equal, the corresponding elements are equal. When you write the sentences to solve this equation, two linear equations are formed.

$y = 3x - 2$

$3 = 2y + x$

This system can be solved using substitution.

$3 = 2y + x$ Second equation

$\boxed{} = 2\left(\boxed{} \right) + x$ Substitute $\boxed{}$ for y.

$\boxed{} = \boxed{} + x$ Distributive Property

$\boxed{} = \boxed{}$ Add 4 to each side.

$\boxed{} = x$ Divide each side by 7.

To find the value for y, substitute 1 for x in either equation.

$y = 3x - 2$ First equation

$y = 3(\boxed{}) - 2$ Substitute $\boxed{}$ for x.

$y = \boxed{}$ Simplify.

The solution is $\boxed{}$.

Your Turn Solve $\begin{bmatrix} y \\ 2x \end{bmatrix} = \begin{bmatrix} 3x - 1 \\ y - 1 \end{bmatrix}$ for x and y.

HOMEWORK ASSIGNMENT

Page(s): _____

Exercises: _____

Operations with Matrices

WHAT YOU'LL LEARN

- Add and subtract matrices.
- Multiply by a matrix scalar.

KEY CONCEPTS

Addition of Matrices
If A and B are two $m \times n$ matrices, then $A + B$ is an $m \times n$ matrix in which each element is the sum of the corresponding elements of A and B.

Subtraction of Matrices
If A and B are two $m \times n$ matrices, then $A - B$ is an $m \times n$ matrix in which each element is the difference of the corresponding elements of A and B.

EXAMPLE Add Matrices

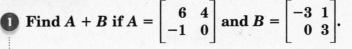

1 Find $A + B$ if $A = \begin{bmatrix} 6 & 4 \\ -1 & 0 \end{bmatrix}$ and $B = \begin{bmatrix} -3 & 1 \\ 0 & 3 \end{bmatrix}$.

$A + B = \begin{bmatrix} 6 & 4 \\ -1 & 0 \end{bmatrix} + \begin{bmatrix} -3 & 1 \\ 0 & 3 \end{bmatrix}$ Definition of matrix addition

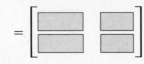

$=$ [ ]   Add corresponding elements.

$=$ [ ]   Simplify.

EXAMPLE Subtract Matrices

2 Find $A - B$ if $A = \begin{bmatrix} 3 & 2 \\ -1 & 0 \end{bmatrix}$ and $B = \begin{bmatrix} -2 & 1 \\ 0 & -1 \end{bmatrix}$.

$A - B = \begin{bmatrix} 3 & 2 \\ -1 & 0 \end{bmatrix} - \begin{bmatrix} -2 & 1 \\ 0 & -1 \end{bmatrix}$ Definition of matrix subtraction

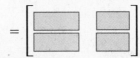

$=$ [ ]   Subtract corresponding elements.

$=$ [ ]   Simplify.

Your Turn

a. Find $A + B$ if
$A = \begin{bmatrix} -2 & 3 \\ 1 & -1 \end{bmatrix}$ and
$B = \begin{bmatrix} 6 & 5 \\ 3 & 0 \end{bmatrix}$.

b. Find $A - B$ if
$A = \begin{bmatrix} 3 & 2 \\ -1 & 0 \end{bmatrix}$ and
$B = \begin{bmatrix} 5 & -2 \\ -1 & 3 \end{bmatrix}$.

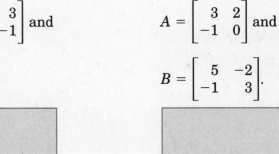

4–2

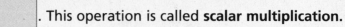

BUILD YOUR VOCABULARY (page 71)

You can multiply any matrix by a constant called a

[   ]. This operation is called **scalar multiplication**.

EXAMPLE Multiply a Matrix by a Scalar

3 If $A = \begin{bmatrix} 2 & 1 \\ -1 & 3 \\ 0 & 5 \end{bmatrix}$, find $2A$.

$2A = \begin{bmatrix} 2 & 1 \\ -1 & 3 \\ 0 & 5 \end{bmatrix}$ Substitution

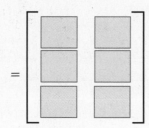

$= \begin{bmatrix} & 1(2) \\ -1(2) & \\ & 5(2) \end{bmatrix}$ Multiply each element by 2.

$= \begin{bmatrix} & \\ & \\ & \end{bmatrix}$ Simplify.

Your Turn If $A = \begin{bmatrix} 2 & 3 & 0 \\ -1 & -5 & 6 \end{bmatrix}$ find $4A$.

KEY CONCEPT

Scalar Multiplication
The product of a scalar k and an $m \times n$ matrix is the matrix in which each element equals k times the corresponding elements of the original matrix.

FOLDABLES Under the tab for Lesson 4-2, write your own example that involves scalar multiplication.
Then perform the multiplication.

HOMEWORK ASSIGNMENT

Page(s):

Exercises:

© Glencoe/McGraw-Hill

76 Glencoe Algebra 2

Multiplying Matrices

What You'll Learn

- Multiply matrices.
- Use the properties of matrix multiplication.

Key Concept

Multiplying Matrices
The element a_{ij} of AB is the sum of the products of the corresponding elements in row i of A and column j of B.

FOLDABLES Under the tab for Lesson 4-3, write an example of multiplying square matrices.

EXAMPLE Dimensions of Matrix Products

1 Determine if the product of $A_{3 \times 4}$ and $B_{4 \times 2}$ is defined. If so, state the dimensions of the product.

$$A \quad \cdot \quad B \quad = \quad AB$$

$$3 \times 4 \quad 4 \times 2 \qquad \boxed{}$$

The inner dimensions are $\boxed{}$ so the matrix product is $\boxed{}$. The dimensions of the product are $\boxed{}$.

Your Turn Determine if the product of $A_{2 \times 3}$ and $B_{2 \times 3}$ is defined. If so, state the dimensions of the product.

EXAMPLE Multiply Square Matrices

2 Find RS if $R = \begin{bmatrix} 3 & 2 \\ -1 & 0 \end{bmatrix}$ and $S = \begin{bmatrix} -2 & 1 \\ 0 & -1 \end{bmatrix}$.

$$RS = \begin{bmatrix} 3 & 2 \\ -1 & 0 \end{bmatrix} \cdot \begin{bmatrix} -2 & 1 \\ 0 & -1 \end{bmatrix}$$

$$= \begin{bmatrix} \boxed{} + \boxed{} & 3(1) + 2(-1) \\ -1(-2) + 0(0) & \boxed{} + \boxed{} \end{bmatrix}$$

$$= \begin{bmatrix} \boxed{} & \boxed{} \\ \boxed{} & \boxed{} \end{bmatrix}$$

Your Turn Find RS if $R = \begin{bmatrix} 3 & -1 \\ 1 & 0 \end{bmatrix}$ and $S = \begin{bmatrix} 2 & 1 \\ -2 & -3 \end{bmatrix}$.

WRITE IT

Is multiplication of matrices commutative? Explain.

_____

_____

_____

_____

_____

EXAMPLE Multiply Matrices with Different Dimensions

3 **CHESS** For each win, a team was awarded 3 points and for each draw a team received 1 point. Find the total number of points for each team. Which team won the tournament?

Team	Wins	Draws
Blue	5	4
Red	6	3
Green	4	5

$RP = \begin{bmatrix} 5 & 4 \\ 6 & 3 \\ 4 & 5 \end{bmatrix} \cdot \begin{bmatrix} 3 \\ 1 \end{bmatrix}$ Write an equation.

$= \begin{bmatrix} \boxed{} + \boxed{} \\ \boxed{} + \boxed{} \\ \boxed{} + \boxed{} \end{bmatrix}$ Multiply columns by rows.

$= \begin{bmatrix} \end{bmatrix}$ Simplify.

The labels for the product matrix are shown. The red team won the championship.

Total Points

Blue

Red

Green

$\begin{bmatrix} \end{bmatrix}$

Your Turn Three players made the points listed below. They scored 1 point for the free-throws, 2 points for the 2-point shots, and 3 points for the 3-point shots. How many points did each player score and who scored the most points?

Player	Free-throws	2-point	3-point
Warton	2	3	2
Bryant	5	1	0
Chris	2	4	5

EXAMPLE Commutative Property

4 Find KL if $K = \begin{bmatrix} -3 & 2 & 2 \\ -1 & -2 & 0 \end{bmatrix}$ and $L = \begin{bmatrix} 1 & -2 \\ 4 & 3 \\ 0 & -1 \end{bmatrix}$.

$KL = \begin{bmatrix} -3 & 2 & 2 \\ -1 & -2 & 0 \end{bmatrix} \cdot \begin{bmatrix} 1 & -2 \\ 4 & 3 \\ 0 & -1 \end{bmatrix}$ Substitution

$= \begin{bmatrix} -3+8+0 & \\ & 2-6+0 \end{bmatrix}$ Multiply.

$= \begin{bmatrix} & \\ & \end{bmatrix}$ Simplify.

Your Turn Find AB if $A = \begin{bmatrix} 0 & 1 \\ 3 & 4 \end{bmatrix}$ and $B = \begin{bmatrix} -2 & 1 \\ 3 & 0 \end{bmatrix}$.

EXAMPLE Distributive Property

5 Find $A(B+C)$ if $A = \begin{bmatrix} -1 & 2 \\ 0 & 1 \end{bmatrix}$, $B = \begin{bmatrix} 1 & 0 \\ 3 & -2 \end{bmatrix}$, and $C = \begin{bmatrix} -3 & 1 \\ -1 & 0 \end{bmatrix}$.

$A(B+C) = \begin{bmatrix} -1 & 2 \\ 0 & 1 \end{bmatrix} \cdot \left(\begin{bmatrix} 1 & 0 \\ 3 & -2 \end{bmatrix} + \begin{bmatrix} -3 & 1 \\ -1 & 0 \end{bmatrix} \right)$ Substitution

$= \begin{bmatrix} -1 & 2 \\ 0 & 1 \end{bmatrix} \cdot \begin{bmatrix} -2 & 1 \\ 2 & -2 \end{bmatrix}$ Add.

$= \begin{bmatrix} & -1-4 \\ & 0-2 \end{bmatrix}$ or $\begin{bmatrix} 6 & \\ 2 & \end{bmatrix}$ Multiply.

Your Turn Find $AB + AC$ if $A = \begin{bmatrix} 3 & -2 \\ 1 & 0 \end{bmatrix}$, $B = \begin{bmatrix} 2 & 0 \\ 1 & -5 \end{bmatrix}$, and $C = \begin{bmatrix} 0 & 1 \\ -6 & 3 \end{bmatrix}$.

HOMEWORK ASSIGNMENT

Page(s):
Exercises:

Transformations with Matrices

WHAT YOU'LL LEARN

- Use matrices to determine the coordinates of a translated or dilated figure.

- Use matrix multiplication to find the coordinates of a reflected or rotated figure.

ORGANIZE IT

Under the tab for Lesson 4-4, write each new Vocabulary Builder word. Then give an example of each word.

BUILD YOUR VOCABULARY (pages 70–71)

Transformations are functions that map points of a **preimage** onto its **image**.

A **translation** occurs when a figure is moved from one location to another without changing its [   ],

[   ], or orientation.

EXAMPLE Translate a Figure

① **Find the coordinates of the vertices of the image of quadrilateral *ABCD* with *A*(−5, −1), *B*(−2, −1), *C*(−1, −4), and *D*(−3, −5) if it is moved 3 units to the right and 4 units up.**

Write the vertex matrix for quadrilateral *ABCD*.

$$\begin{bmatrix} & & -1 & \\ -1 & -1 & & -5 \end{bmatrix}$$

To translate the quadrilateral 3 units to the right, add 3 to each *x*-coordinate. To translate the figure 4 units up, add 4 to each *y*-coordinate. This can be done by adding the translation matrix to the vertex matrix.

Vertex Matrix   Translation
of *ABCD*   Matrix

$$\begin{bmatrix} -5 & -2 & -1 & -3 \\ -1 & -1 & -4 & -5 \end{bmatrix} + \boxed{}$$

$$= \boxed{}$$

The coordinates of *A'B'C'D'* are

A' [   ], *B'* [   ],

C' [   ], *D'* [   ].

By graphing the preimage and the image, you find that the coordinates of *A'B'C'D'* are correct.

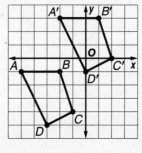

Your Turn Find the coordinates of the vertices of the image of quadrilateral *HIJK* with *H*(2, 3), *I*(3, −1), *J*(−1, −3), and *K*(−2, 5) if it is moved 2 units to the left and 2 units up.

BUILD YOUR VOCABULARY (pages 70–71)

A **reflection** occurs when every point of a figure is mapped

to a corresponding image across [ ]

using a *reflection matrix*.

A **rotation** occurs when a figure is moved around a

[ ] , usually the [ ] .

EXAMPLE Reflection

KEY CONCEPT

Reflection Matrices

x-axis $\begin{bmatrix} 1 & 0 \\ 0 & -1 \end{bmatrix}$

y-axis $\begin{bmatrix} -1 & 0 \\ 0 & 1 \end{bmatrix}$

line $y = x$ $\begin{bmatrix} 0 & 1 \\ 1 & 0 \end{bmatrix}$

2 Find the coordinates of the vertices of the image of pentagon *PENTA* with *P*(−3, 1), *E*(0, −1), *N*(−1, −3), *T*(−3, −4), and *A*(−4, −1) after a reflection across the *x*-axis.

Write the ordered pairs as a vertex matrix. Then [ ] the vertex matrix by the [ ] matrix for the *x*-axis.

$$\begin{bmatrix} \end{bmatrix} \cdot \begin{bmatrix} -3 & 0 & -1 & -3 & -4 \\ 1 & -1 & -3 & -4 & -1 \end{bmatrix}$$

$$= \begin{bmatrix} -3 & & -1 & & \\ & 1 & & 4 & 1 \end{bmatrix}$$

The coordinates of the vertices of *P'E'N'T'A'* are *P'*(−3, −1), *E'*(0, 1), *N'*(−1, 3), *T'*(−3, 4), and *A'*(−4, 1). The graph of the preimage and image shows that the coordinates of *P'E'N'T'A'* are correct.

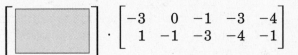

Your Turn Find the coordinates of the vertices of the image of pentagon *PENTA* with $P(-5, 0)$, $E(-3, 3)$, $N(1, 2)$, $T(1, -1)$, and $A(-4, -2)$ after a reflection across the line $y = x$.

EXAMPLE Rotation

KEY CONCEPT

Reflection Matrices

90°: $\begin{bmatrix} 0 & -1 \\ 1 & 0 \end{bmatrix}$

180°: $\begin{bmatrix} -1 & 0 \\ 0 & -1 \end{bmatrix}$

270°: $\begin{bmatrix} 0 & 1 \\ -1 & 0 \end{bmatrix}$

④ **Find the coordinates of the vertices of the image of △*DEF* with $D(4, 3)$, $E(1, 1)$, and $F(2, 5)$ after it is rotated 90° counterclockwise about the origin.**

Write the ordered pairs in a vertex matrix. Then multiply the vertex matrix by the rotation matrix.

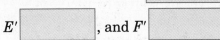 $\cdot \begin{bmatrix} 4 & 1 & 2 \\ 3 & 1 & 5 \end{bmatrix} = \begin{bmatrix} & & -5 \\ 4 & 1 & \end{bmatrix}$

The coordinates of the vertices of triangle $D'E'F'$ are D' [   ], E' [   ], and F' [   ].
The graph of the preimage and image show that the coordinates of $D'E'F'$ are correct.

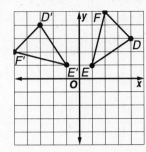

Your Turn Find the coordinates of the vertices of the image of △*TRI* with $T(-1, 2)$, $R(-3, 0)$ and $I(-2, -2)$ after it is rotated 180° counterclockwise about the origin.

HOMEWORK ASSIGNMENT

Page(s): _____
Exercises: _____

_____
_____

Determinants

What You'll Learn

- Evaluate the determinant of a 2 × 2 matrix.

- Evaluate the determinant of a 3 × 3 matrix.

Key Concept

Second-Order Determinant The value of a second-order determinant is found by calculating the difference of the products of the two diagonals.

EXAMPLE Second-Order Determinant

1 Find the value of $\begin{vmatrix} -6 & 7 \\ -9 & 3 \end{vmatrix}$.

$$\begin{vmatrix} -6 & 7 \\ -9 & 3 \end{vmatrix} = \boxed{} - \boxed{} \qquad \text{Definition of determinant}$$

$$= \boxed{} + \boxed{} \qquad \text{Multiply.}$$

$$= \boxed{} \qquad \text{Simplify.}$$

EXAMPLE Expansion by Minors

2 Evaluate $\begin{vmatrix} 1 & 0 & -1 \\ 2 & -1 & 3 \\ 4 & -2 & -3 \end{vmatrix}$ using expansion by minors.

Decide which row of elements to use for the expansion. For this example, let's use the first row.

$$\begin{vmatrix} 1 & 0 & -1 \\ 2 & -1 & 3 \\ 4 & -2 & -3 \end{vmatrix}$$

$$= 1 \begin{vmatrix} -1 & \boxed{} \\ \boxed{} & -3 \end{vmatrix} - 0 \begin{vmatrix} 2 & 3 \\ 4 & -3 \end{vmatrix} + (-1) \begin{vmatrix} \boxed{} & -1 \\ 4 & \boxed{} \end{vmatrix}$$

$$= 1(3 - (\boxed{})) - 0(\boxed{} - 12) - 1(-4 - (\boxed{}))$$

$$= 1(\boxed{}) - \boxed{} - 1(\boxed{})$$

$$= \boxed{}$$

EXAMPLE Use Diagonals

3 Evaluate $\begin{vmatrix} 3 & -2 & -1 \\ 2 & -1 & 0 \\ 1 & 2 & -3 \end{vmatrix}$ using diagonals.

STEP 1 Rewrite the first 2 columns to the right of the determinant.

$\begin{vmatrix} 3 & -2 & -1 \\ 2 & -1 & 0 \\ 1 & 2 & -3 \end{vmatrix}$

STEP 2 Find the product of the elements of the diagonals.

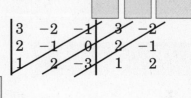

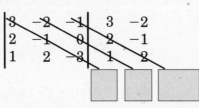

STEP 3 Add the bottom products and subtract the top products.

 =

The value of the determinant is ___.

Your Turn Evaluate each determinant.

a. $\begin{vmatrix} 3 & 2 \\ 1 & 0 \end{vmatrix}$

b. $\begin{bmatrix} -3 & 2 & 0 \\ 1 & 4 & 1 \\ 2 & 3 & 6 \end{bmatrix}$

c. $\begin{vmatrix} 2 & -3 & -1 \\ 5 & 0 & -2 \\ 1 & 2 & 5 \end{vmatrix}$

HOMEWORK ASSIGNMENT

Page(s):

Exercises:

Cramer's Rule

WHAT YOU'LL LEARN

- Solve systems of two linear equations by using Cramer's Rule.

- Solve systems of three linear equations by using Cramer's Rule.

KEY CONCEPT

Cramer's Rule for Two Variables The solution of the system of linear equations $ax + by = e$ and $cx + dy = f$ is (x, y), where

$$x = \frac{\begin{vmatrix} e & b \\ f & d \end{vmatrix}}{\begin{vmatrix} a & b \\ c & d \end{vmatrix}} , \quad y = \frac{\begin{vmatrix} a & e \\ c & f \end{vmatrix}}{\begin{vmatrix} a & b \\ c & d \end{vmatrix}} ,$$

and $\begin{vmatrix} a & b \\ c & d \end{vmatrix} \neq 0$.

FOLDABLES Under the tab for Lesson 4-6, write this rule.

BUILD YOUR VOCABULARY (page 70)

Cramer's Rule uses determinants to solve systems of equations.

EXAMPLE System of Two Equations

1 Use Cramer's Rule to solve $5x + 4y = 28$ and $3x - 2y = 8$.

$$x = \frac{\begin{vmatrix} e & b \\ f & d \end{vmatrix}}{\begin{vmatrix} a & b \\ c & d \end{vmatrix}} \qquad \text{Cramer's Rule} \qquad y = \frac{\begin{vmatrix} a & e \\ c & f \end{vmatrix}}{\begin{vmatrix} a & b \\ c & d \end{vmatrix}}$$

$$= \frac{\begin{vmatrix} & \\ 8 & \end{vmatrix}}{\begin{vmatrix} 5 & 4 \\ 3 & \end{vmatrix}} \qquad \begin{array}{c} a = 5, b = 4, \\ c = 3, d = -2, \\ e = 28, \text{ and} \\ f = 8 \end{array} \qquad = \frac{\begin{vmatrix} & 28 \\ 3 & \end{vmatrix}}{\begin{vmatrix} 5 & 4 \\ 3 & \end{vmatrix}}$$

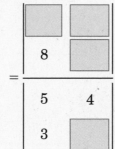 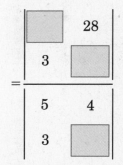

$$= \frac{\boxed{} - \boxed{}}{5(-2) - 3(4)} \qquad \text{Evaluate.} \qquad = \frac{5(8) - 3(28)}{5(-2) - 3(4)}$$

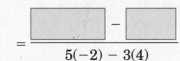

$$= \boxed{} \text{ or } \boxed{} \qquad \text{Simplify.} \qquad = \boxed{} \text{ or } \boxed{}$$

The solution is $\boxed{}$.

Your Turn Use Cramer's Rule to solve $3x + 2y = 1$ and $2x - 5y = -12$.

KEY CONCEPT

The solution of the system whose equations are
$ax + by + cz = j$
$dx + ey + fz = k$
$gx + hy + iz = l$
is (x, y, z), where

$$x = \frac{\begin{vmatrix} j & b & c \\ k & e & f \\ l & h & i \end{vmatrix}}{\begin{vmatrix} a & b & c \\ d & e & f \\ g & h & i \end{vmatrix}},$$

$$y = \frac{\begin{vmatrix} a & j & c \\ d & k & f \\ g & l & i \end{vmatrix}}{\begin{vmatrix} a & b & c \\ d & e & f \\ g & h & i \end{vmatrix}},$$

$$z = \frac{\begin{vmatrix} a & b & j \\ d & e & k \\ g & h & l \end{vmatrix}}{\begin{vmatrix} a & b & c \\ d & e & f \\ g & h & i \end{vmatrix}}, \text{ and}$$

and $\begin{vmatrix} a & b & c \\ d & e & f \\ g & h & i \end{vmatrix} \neq 0.$

HOMEWORK ASSIGNMENT

Page(s): _____

Exercises: _____

EXAMPLE System of Three Equations

2 Use Cramer's Rule to solve the system of equations.

$2x - 3y + z = 5$
$x + 2y + z = -1$
$x - 3y + 2z = 1$

$$x = \frac{\begin{vmatrix} j & b & c \\ k & e & f \\ l & h & i \end{vmatrix}}{\begin{vmatrix} a & b & c \\ d & e & f \\ g & h & i \end{vmatrix}} = \frac{\begin{vmatrix} 5 & -3 & 1 \\ -1 & 2 & 1 \\ 1 & -3 & 2 \end{vmatrix}}{\begin{vmatrix} 2 & -3 & 1 \\ 1 & 2 & 1 \\ 1 & -3 & 2 \end{vmatrix}}$$

$$y = \frac{\begin{vmatrix} a & j & c \\ d & k & f \\ g & l & i \end{vmatrix}}{\begin{vmatrix} a & b & c \\ d & e & f \\ g & h & i \end{vmatrix}} = \frac{\begin{vmatrix} 2 & 5 & 1 \\ 1 & -1 & 1 \\ 1 & 1 & 2 \end{vmatrix}}{\begin{vmatrix} 2 & -3 & 1 \\ 1 & 2 & 1 \\ 1 & -3 & 2 \end{vmatrix}}$$

$$z = \frac{\begin{vmatrix} a & b & j \\ d & e & k \\ g & h & l \end{vmatrix}}{\begin{vmatrix} a & b & c \\ d & e & f \\ g & h & i \end{vmatrix}} = \frac{\begin{vmatrix} 2 & -3 & 5 \\ 1 & 2 & -1 \\ 1 & -3 & 1 \end{vmatrix}}{\begin{vmatrix} 2 & -3 & 1 \\ 1 & 2 & 1 \\ 1 & -3 & 2 \end{vmatrix}}$$

Use a calculator to evaluate each determinant.

$x = \dfrac{}{}$ or $\dfrac{9}{4}$ $y = \dfrac{-9}{12}$ or $-\dfrac{}{}$ $z = \dfrac{}{}$ or $-\dfrac{7}{4}$

The solution is _____.

Your Turn Use Cramer's Rule to solve the system of equations.

$2x + y + z = -3$
$-3x + 2y - z = 5$
$x - y + 3z = 1$

WHAT YOU'LL LEARN

- Determine whether two matrices are inverses.
- Find the inverse of a 2 × 2 matrix.

BUILD YOUR VOCABULARY (page 70)

The **identity matrix** is a square matrix that, when multiplied by another matrix, equals that same matrix.

Two $n \times n$ matrices are **inverses** of each other if their

[_____] is the [_____].

KEY CONCEPTS

Identity Matrix for Multiplication The identity matrix for multiplication I is a square matrix with 1 for every element of the main diagonal, from upper left to lower right, and 0 in all other positions. For any square matrix A of the same dimensions as I, $A \cdot I = I \cdot A = A$.

Inverse of a 2 × 2 matrix
The inverse of Matrix A
$= \begin{vmatrix} a & b \\ c & d \end{vmatrix}$ is $A^{-1} =$
$\dfrac{1}{ad - bc} \begin{vmatrix} d & -b \\ -c & a \end{vmatrix}$, where $ad - bc \neq 0$.

EXAMPLE Verify Inverse Matrices

1 Determine whether $X = \begin{bmatrix} 3 & -2 \\ -1 & 1 \end{bmatrix}$ and $Y = \begin{bmatrix} 1 & 2 \\ 1 & 3 \end{bmatrix}$ are inverses.

Find $X \cdot Y$.

$X \cdot Y = \begin{bmatrix} 3 & -2 \\ -1 & 1 \end{bmatrix} \cdot \begin{bmatrix} 1 & 2 \\ 1 & 3 \end{bmatrix}$

$= \begin{bmatrix} \boxed{} & 6 - 6 \\ -1 + 1 & \boxed{} \end{bmatrix}$ or $\begin{bmatrix} \boxed{} & \boxed{} \\ \boxed{} & \boxed{} \end{bmatrix}$

Find $Y \cdot X$.

$Y \cdot X = \begin{bmatrix} 1 & 2 \\ 1 & 3 \end{bmatrix} \cdot \begin{bmatrix} 3 & -2 \\ -1 & 1 \end{bmatrix}$

$= \begin{bmatrix} 3 - 2 & \boxed{} \\ 3 - 3 & \boxed{} \end{bmatrix}$ or $\begin{bmatrix} \boxed{} & \boxed{} \\ \boxed{} & \boxed{} \end{bmatrix}$

Since $X \cdot Y = Y \cdot X = I$, X and Y are inverses.

Your Turn Determine whether each pair of matrices are inverses.

a. $A = \begin{bmatrix} -2 & 3 \\ 1 & 0 \end{bmatrix}$ and $B = \begin{bmatrix} -2 & 2 \\ 1 & 1 \end{bmatrix}$

b. $C = \begin{bmatrix} 3 & 1 \\ 2 & 1 \end{bmatrix}$ and $D = \begin{bmatrix} 1 & -1 \\ -2 & 3 \end{bmatrix}$

EXAMPLE Find the Inverse of a Matrix

② **Find the inverse of each matrix, if it exists.**

a. $S = \begin{bmatrix} -1 & 0 \\ 8 & -2 \end{bmatrix}$

Find the value of the determinant.

$$\begin{bmatrix} -1 & 0 \\ 8 & 2 \end{bmatrix} = 2 - \boxed{} = \boxed{}$$

Since the determinant is not equal to 0, S^{-1} exists.

$$S^{-1} = \frac{1}{ad - bc} \begin{bmatrix} d & -b \\ -c & a \end{bmatrix}$$

$$= \frac{1}{-1(-2) - \boxed{}(0)} \begin{bmatrix} \boxed{} & 0 \\ -8 & \boxed{} \end{bmatrix}$$

$$= \frac{1}{\boxed{}} \begin{bmatrix} -2 & 0 \\ -8 & -1 \end{bmatrix} \text{ or } \begin{bmatrix} \boxed{} & \boxed{} \\ \boxed{} & -\frac{1}{2} \end{bmatrix}$$

FOLDABLES

ORGANIZE IT

Under the tab for Lesson 4-7, write your own 2 × 2 matrix. Then find the inverse of the matrix, if it exists.

4-1 Introduction
4-2 Operations
4-3 Multiplying Matrices
4-4 Transformations
4-5 Determinants
4-6 Cramer's Rule
4-7 Identity
4-8 Using Matrices

b. $T = \begin{bmatrix} -4 & 6 \\ -2 & 3 \end{bmatrix}$

Find the value of the determinant.

$$\begin{vmatrix} -4 & 6 \\ -2 & 3 \end{vmatrix} = \boxed{} + \boxed{} = \boxed{}$$

Since the determinant equals 0, T^{-1} does not exist.

Your Turn Find the inverse of each matrix, if it exists.

a. $A = \begin{bmatrix} 2 & 1 \\ 6 & 3 \end{bmatrix}$

b. $B = \begin{bmatrix} 6 & -2 \\ 7 & -2 \end{bmatrix}$

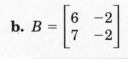

HOMEWORK ASSIGNMENT

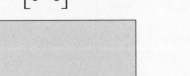

Page(s):

Exercises:

WHAT YOU'LL LEARN

- Write matrix equations for systems of equations.

- Solve systems of equations using matrix equations.

FOLDABLES

ORGANIZE IT

Under the tab for Lesson 4-8, write a matrix equation for this system of equations.
$x + 2y = 5$
$3x + 5y = 14$

BUILD YOUR VOCABULARY (page 71)

A system of equations can be written with [   ] expressing the system of equations as a **matrix equation**.

EXAMPLE Two-Variable Matrix Equation

1 **Write a matrix equation for the system of equations.**

$x + 3y = 3$
$x + 2y = 7$

Determine the coefficient, variable, and constant matrices.

$x + 3y = 3$
$x + 2y = 7$

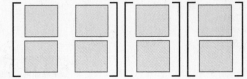

Write the matrix equation.

$$A \qquad \cdot \qquad X \qquad = \qquad B$$

EXAMPLE Solve a System of Equations

2 **Use a matrix equation to solve the system of equations.**

$5x + 3y = 13$
$4x + 7y = -8$

The matrix equation is

when $A =$ [   ], $X =$ [   ], and $B =$ [   ].

STEP 1 Find the inverse of the coefficient matrix.

$$A^{-1} = \frac{\boxed{}}{\boxed{} - \boxed{}} \begin{bmatrix} 7 & -3 \\ -4 & 5 \end{bmatrix} \text{ or } \boxed{} \begin{bmatrix} 7 & -3 \\ -4 & 5 \end{bmatrix}$$

STEP 2 Multiply each side of the matrix equation by the inverse matrix.

$$\frac{1}{23}\begin{bmatrix} 7 & -3 \\ -4 & 5 \end{bmatrix} \cdot \begin{bmatrix} 5 & 3 \\ 4 & 7 \end{bmatrix} \cdot \begin{bmatrix} x \\ y \end{bmatrix} = \frac{1}{23}\begin{bmatrix} 7 & -3 \\ -4 & 5 \end{bmatrix} \cdot \begin{bmatrix} 13 \\ -8 \end{bmatrix}$$

$$\boxed{} \cdot \boxed{} = \frac{1}{23} \boxed{}$$

$$\boxed{} = \boxed{}$$

The solution is $\boxed{}$.

Your Turn

a. Write a matrix equation for the system of equations.
$x - 2y = 6$ and $3x + 4y = 7$

b. Use a matrix equation to solve the system of equations.
$3x + 4y = -10$ and $x - 2y = 10$

HOMEWORK ASSIGNMENT

Page(s): _____

Exercises: _____

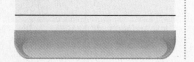

BRINGING IT ALL TOGETHER

STUDY GUIDE

FOLDABLES™	VOCABULARY PUZZLEMAKER	**BUILD YOUR VOCABULARY**
Use your **Chapter 4 Foldable** to help you study for your chapter test.	To make a crossword puzzle, word search, or jumble puzzle of the vocabulary words in Chapter 4, go to: www.glencoe.com/sec/math/t_resources/free/index.php	You can use your completed **Vocabulary Builder** (pages 70–71) to help you solve the puzzle.

4-1
Introduction to Matrices

Match each matrix with its dimensions.

1. $\begin{bmatrix} 3 & 2 & 5 \\ -1 & 0 & 6 \end{bmatrix}$ [ ]

2. $\begin{bmatrix} 30 & -84 \end{bmatrix}$ [ ]

3. $\begin{bmatrix} 0 & 3 \\ 1 & -2 \end{bmatrix}$ [ ]

4. $\begin{bmatrix} 4 & 0 \\ -1 & 2 \\ 6 & 1 \end{bmatrix}$ [ ]

a. 3×2

b. 2×3

c. 2×2

d. 1×2

5. Write a system of equations that you could use to solve the following matrix equation for x, y, z.

$$\begin{bmatrix} 3x \\ x + y \\ y - z \end{bmatrix} = \begin{bmatrix} -9 \\ 5 \\ 6 \end{bmatrix}$$

4-2
Operations with Matrices

6. Use $M = \begin{bmatrix} 3 & 0 & 2 \\ 2 & -1 & 4 \end{bmatrix}$ and $N = \begin{bmatrix} -2 & 5 & -4 \\ 3 & 1 & 0 \end{bmatrix}$ to find $2M + 3N$.

4-3

Multiplying Matrices

Determine whether each indicated matrix product is defined. If so, state the dimensions of the product. If not, write *undefined*.

7. $M_{3 \times 2}$ and $N_{2 \times 3}$ *MN*:

8. $M_{1 \times 2}$ and $N_{1 \times 2}$ *MN*:

9. $M_{4 \times 1}$ and $N_{1 \times 4}$ *MN*:

10. Find the product, if possible.

$$\begin{bmatrix} 2 & 0 & 3 \\ 1 & -1 & 5 \end{bmatrix} \cdot \begin{bmatrix} 1 & 3 \\ 5 & 2 \\ -3 & 0 \end{bmatrix}$$

4-4

Transformations with Matrices

Refer to quadrilateral *ABCD* shown.

11. Write the vertex matrix for the quadrilateral *ABCD*.

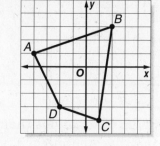

12. Write the vertex matrix that represents the position of the quadrilateral $A'B'C'D'$ that results when quadrilateral *ABCD* is translated 3 units to the right and 2 units down.

4-5

Determinants

13 Find the value of $\begin{vmatrix} 8 & 3 \\ 2 & -1 \end{vmatrix}$.

14. Evaluate $\begin{vmatrix} 3 & 12 & -1 \\ 10 & 9 & 0 \\ -5 & 6 & -2 \end{vmatrix}$ using expansion by minors.

4-6
Cramer's Rule

15. The two sides of an angle are contained in the lines whose equations are $3x + y = 5$ and $2x + 3y = 8$. Find the coordinates of the vertex of the angle.

16. Use Cramer's Rule to solve the system of equations.

$2x + 5y + 3z = 10$
$3x - y + 4z = 8$
$5x - 2y + 7z = 12$

4-7
Identity and Inverse Matrices

Indicate whether each of the following statements is *true* or *false*.

17. Every element of an identity matrix is 1.

18. There is a 3×2 identity matrix.

19. If M is a matrix, M^{-1} represents the reciprocal of M.

20. Every square matrix has an inverse.

21. Determine whether $A = \begin{bmatrix} 1 & -2 \\ -3 & 7 \end{bmatrix}$ and

$B = \begin{bmatrix} 7 & 2 \\ 3 & 1 \end{bmatrix}$ are inverses.

4-8
Using Matrices to Solve Systems of Equations

22. Write a matrix equation for the following system of equations.

$3x + 5y = 10$
$2x - 4y = -7$

23. Solve the system of equations $4x - 5y = 4$ and $2x - y = 8$ by using inverse matrices.

ARE YOU READY FOR THE CHAPTER TEST?

Check the one that applies. Suggestions to help you study are given with each item.

☐ **I completed the review of all or most lessons without using my notes or asking for help.**

- You are probably ready for the Chapter Test.

- You may want to take the Chapter 4 Practice Test on page 215 of your textbook as a final check.

☐ **I used my Foldable or Study Notebook to complete the review of all or most lessons.**

- You should complete the Chapter 4 Study Guide and Review on pages 209–214 of your textbook.

- If you are unsure of any concepts or skills, refer back to the specific lesson(s).

- You may also want to take the Chapter 4 Practice Test on page 215.

☐ **I asked for help from someone else to complete the review of all or most lessons.**

- You should review the examples and concepts in your Study Notebook and Chapter 4 Foldable.

- Then complete the Chapter 4 Study Guide and Review on pages 209–214 of your textbook.

- If you are unsure of any concepts or skills, refer back to the specific lesson(s).

- You may also want to take the Chapter 4 Practice Test on page 215 of your textbook.

Student Signature	Parent/Guardian Signature

Teacher Signature

CHAPTER 5

Polynomials

 Use the instructions below to make a Foldable to help you organize your notes as you study the chapter. You will see Foldable reminders in the margin of this Interactive Study Notebook to help you in taking notes.

Begin with four sheets of grid paper.

STEP 1 **Fold and Cut**
Fold in half along the width. On the first two sheets, cut along the fold at the ends. On the second two sheets, cut in the center of the fold as shown.

First Sheets   Second Sheets

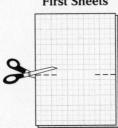 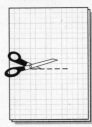

STEP 2 **Fold and Label**
Insert first sheets through second sheets and align folds. Label pages with lesson numbers.

 NOTE-TAKING TIP: When you take notes, you may wish to use a highlighting marker to emphasize important concepts.

BUILD YOUR VOCABULARY

This is an alphabetical list of new vocabulary terms you will learn in Chapter 5.
As you complete the study notes for the chapter, you will see Build Your
Vocabulary reminders to complete each term's definition or description on
these pages. Remember to add the textbook page number in the second
column for reference when you study.

Vocabulary Term	Found on Page	Definition	Description or Example
binomial			
coefficient [KOH-uh-FIH-shuhnt]			
complex conjugates [KAHN-jih-guht]			
complex number			
degree			
extraneous solution [ehk-STRAY-nee-uhs]			
FOIL method			
imaginary unit			
like radical expressions			
like terms			

Vocabulary Term	Found on Page	Definition	Description or Example
monomial			
*n*th root			
polynomial			
power			
principal root			
pure imaginary number			
radical equation			
radical inequality			
rationalizing the denominator			
synthetic division [sihn-THEH-tihk]			
trinomial			

Monomials

© Glencoe/McGraw-Hill

WHAT YOU'LL LEARN

- Multiply and divide monomials.

- Use expressions written in scientific notation.

BUILD YOUR VOCABULARY (pages 96–97)

A **monomial** is an expression that is a number, a variable, or the product of a number and one or more variables.

The numerical factor of a monomial is the **coefficient** of the variable(s). The **degree** of a monomial is the sum of the exponents of its variables. A **power** is an expression of the form x^n.

EXAMPLE Simplify Expressions with Multiplication

1 Simplify $(-2a^3b)(-5ab^4)$.

$(-2a^3b)(-5ab^4)$

$= (-2 \cdot a \cdot a \cdot a \cdot b)$ [_____]   Definition of exponents

$= -2(-5) \cdot a \cdot a \cdot a \cdot a \cdot b \cdot b \cdot b \cdot b \cdot b$ Commutative Property

$= 10a^4b^5$ Definition of exponents

EXAMPLE Simplify Expressions with Division

2 Simplify $\dfrac{s^2}{s^{10}}$. Assume that $s \neq 0$.

$\dfrac{s^2}{s^{10}} =$ [_____]   Subtract exponents.

$=$ [_____] or [_____]   A simplified expression cannot contain negative exponents.

EXAMPLE Simplify Expressions with Powers

3 Simplify each expression.

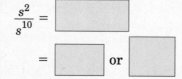

a. $(-3c^2d^5)^3$

$(-3c^2d^5)^3 =$ [_____]

$=$ [_____]

b. $\left(\dfrac{x}{3}\right)^{-4}$

$\left(\dfrac{x}{3}\right)^{-4} = \left(\dfrac{3}{x}\right)^4$

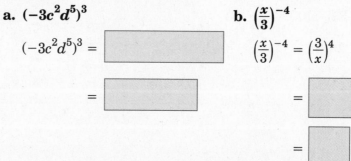

$=$ [_____]

$=$ [_____]

KEY CONCEPTS

Negative Exponents
For any real number $a \neq 0$ and any integer n, $a^{-n} = \dfrac{1}{a^n}$ and $\dfrac{1}{a^{-n}} = a^n$.

Product of Powers
For any real number a and integers m and n, $a^m \cdot a^n = a^{m+n}$.

Quotient of Powers
For any real number $a \neq 0$, and integers m and n, $\dfrac{a^m}{a^n} = a^{m-n}$.

Properties of Powers
Suppose a and b are real numbers and m and n are integers. Then the following properties hold.

Power of a Power:
$(a^m)^n = a^{mn}$

Power of a Product:
$(ab)^m = a^m b^m$

Power of a Quotient:
$\left(\dfrac{a}{b}\right)^n = \dfrac{a^n}{b^n}$, $b \neq 0$ and $\left(\dfrac{a}{b}\right)^{-n} = \left(\dfrac{b}{a}\right)^n$ or $\dfrac{b^n}{a^n}$, $a \neq 0$, $b \neq 0$

 Simplify each expression.

a. $(-3x^2y)(5x^3y^5)$

b. $\dfrac{x^3}{x^7}$

c. $(x^3)^5$

d. $(-2x^2y^3)^5$

e. $\left(\dfrac{-3x^2}{y^3}\right)^3$

f. $\left(\dfrac{y}{2}\right)^{-3}$

EXAMPLE Simplify Expressions Using Several Properties

4 Simplify $\left(\dfrac{-3a^{5y}}{a^{6y}b^4}\right)^5$.

Simplify the fraction before raising to the fifth power.

$\left(\dfrac{-3a^{5y}}{a^{6y}b^4}\right)^5 = \left(\dfrac{-3a^{5y-6y}}{b^4}\right)^5$

$= \left(\dfrac{-3a^{-y}}{b^4}\right)^5$

$=$ [   ] or [   ]

 Simplify $\left(\dfrac{-2a^{3n}}{a^{2n}b^5}\right)^3$.

EXAMPLE Express Numbers in Scientific Notation

5 Express each number in scientific notation.

a. **4,560,000**

$4,560,000 = 4.56 \times 1,000,000$ $1 \le 4.56 < 10$

$=$ [   ]   Write 1,000,000 as a power of 10.

b. **0.000092**

$0.000092 =$ [   ] $\times 0.00001$ $1 \le$ [   ] < 10

$=$ [   ] or [   ]

WRITE IT

Write how you read the expression a^3. Then do the same for z^4.

_____

_____

_____

_____

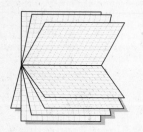

EXAMPLE Multiply Numbers in Scientific Notation

6 **Evaluate. Express the result in scientific notation.**

a. $(5 \times 10^3)(7 \times 10^8)$

$(5 \times 10^3)(7 \times 10^8)$

$= \boxed{} \times \boxed{}$ Associative and Commutative Properties

$= 35 \times 10^{11}$ Simplify.

$= \boxed{}$ $1 \le 3.5 < 10$

b. $(1.8 \times 10^{-4})(4 \times 10^7)$

$(1.8 \times 10^{-4})(4 \times 10^7)$

$= (1.8 \cdot 4) \times (10^{-4} \cdot 10^7)$ Associative and Commutative Properties

$= \boxed{}$ Simplify.

Your Turn **Evaluate. Express the result in scientific notation.**

a. 0.000127

b. $(1.2 \times 10^{-3})(5 \times 10^5)$

EXAMPLE Divide Numbers in Scientific Notation

7 **BIOLOGY** There are about 5×10^6 red blood cells in one milliliter of blood. A certain blood sample contains 8.32×10^6 red blood cells. About how many milliliters of blood are in the sample?

Divide the number of red blood cells in the sample by the number of red blood cells in 1 milliliter of blood.

$$\frac{\boxed{}}{5 \times 10^6} \longleftarrow \text{number of red blood cells in sample}$$
$$ \longleftarrow \text{number of red blood cells in 1 milliliter}$$

$= \boxed{}$ milliliters

Your Turn A petri dish contains 3.6×10^5 germs. A half hour later, there are 7.2×10^7. How many times as great is the amount a half hour later?

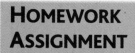

Polynomials

WHAT YOU'LL LEARN

• Add and subtract polynomials.

• Multiply polynomials.

BUILD YOUR VOCABULARY (pages 96–97)

A **polynomial** is a monomial or a sum of monomials.

The polynomial $x^2 + 3x + 1$ is a **trinomial** because it has three unlike terms.

A polynomial such as $xy + z^3$ is a **binomial** because it has two unlike terms.

EXAMPLE Degree of a Polynomial

1 Determine whether $c^4 - 4\sqrt{c} + 18$ is a polynomial. If it is a polynomial, state the degree of the polynomial.

This expression [   ] a polynomial because [   ] is not a monomial.

EXAMPLE Subtract and Simplify

2 Simplify $(2a^3 + 5a - 7) - (a^3 - 3a + 2)$.

$(2a^3 + 5a - 7) - (a^3 - 3a + 2)$

$= 2a^3 + 5a - 7 - a^3 + 3a - 2$

$= (2a^3 - a^3) +$ [   ] $+$ [   ]

$=$ [   ]

REMEMBER IT

The prefix *bi-* means two, so you can remember that a binomial has two unlike terms.

The prefix *tri-* means three, so you can remember that a trinomial has three unlike terms.

Your Turn Determine whether each expression is a polynomial. If it is a polynomial, state the degree of the polynomial.

a. $\frac{1}{2}a^2b^3 + 3c^5$

[   ]

b. $\sqrt{c} + 2$

[   ]

c. Simplify $(3x^2 + 2x - 3) - (4x^2 + x - 5)$.

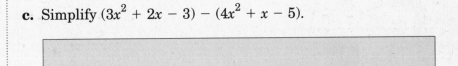

EXAMPLE Multiply and Simplify

3 Find $-y(4y^2 + 2y - 3)$.

$-y(4y^2 + 2y - 3)$

$= -y(4y^2) - y(2y) - y(-3)$ Distributive Property

$=$ [ ]   Multiply the monomials.

EXAMPLE Multiply Two Binomials

4 Find $(2p + 3)(4p + 1)$.

$(2p + 3)(4p + 1)$

$=$ [ ] $+$ [ ] $+$ [ ] $+$ [ ]

    First terms   Outer terms   Inner terms   Last terms

$=$ [ ]   Multiply monomials and add like terms.

EXAMPLE Multiply Polynomials

5 Find $(a^2 + 3a - 4)(a + 2)$

$(a^2 + 3a - 4)(a + 2)$

$= a^2(a + 2) + 3a(a + 2) - 4(a + 2)$ Distributive Property

$= a^2 \cdot a + a^2 \cdot 2 +$ [ ] $+$

[ ] $-$ [ ] $-$ [ ]   Distributive Property

$= a^3 + 2a^2 + 3a^2 + 6a - 4a - 8$ Multiply monomials.

$=$ [ ]   Combine like terms.

Your Turn Find each product.

a. $-x(3x^3 - 2x + 5)$

b. $(3p + 2)(5p + 1)$

c. $(x^2 + 3x - 2)(x + 4)$

Dividing Polynomials

WHAT YOU'LL LEARN

- Divide polynomials using long division.

- Divide polynomials using synthetic division.

FOLDABLES

ORGANIZE IT

In your notes, explain how to write the answer to a long division problem that has a quotient and a remainder.

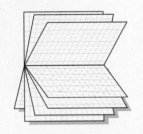

EXAMPLE Division Algorithm

1 Use long division to find $(x^2 - 2x - 15) \div (x - 5)$.

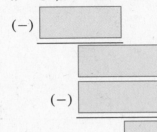

$x(x - 5) = x^2 - 5x$

$-2x - (-5x) = 3x$

$3(x - 5) = 3x - 15$

The quotient is [ ].

EXAMPLE Synthetic Division

2 Use synthetic division to find $(x^3 - 4x^2 + 6x - 4) \div (x - 2)$.

STEP 1 With the terms of the dividend in descending order by degree, write just the coefficients.

$x^3 - 4x^2 + 6x - 4$

$1 \quad -4$

STEPS 2 & 3 Write the constant of the divisor $x - r$, 2, to the left. Bring down the first coefficient, 1.

Multiply the first coefficient by r, $1 \cdot 2 = 2$. Write the product under the next coefficient and add.

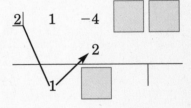

STEP 4 Multiply this sum by r, 2. Write the product under the next coefficient and add.

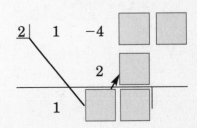

STEP 5 Multiply this sum by r, 2. Write the product under the next coefficient and add.

The numbers along the bottom are the coefficients of the quotient. So, the quotient is $x^2 - 2x + 2$.

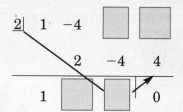

EXAMPLE Divisor with First Coefficient Other than 1

❸ Use synthetic division to find $(4y^4 - 5y^2 + 2y + 4) \div (2y - 1)$.

Use division to rewrite the divisor so it has a first coefficient of 1.

$$\frac{4y^4 - 5y^2 + 2y + 4}{2y - 1} = \frac{(4y^4 - 5y^2 + 2y + 4) \div 2}{(2y - 1) \div 2}$$

$$= \frac{\boxed{}}{y - \frac{1}{2}}$$

The numerator does not have a y^3 term. So use a coefficient of 0.

$$\begin{array}{r|rrrrr} \frac{1}{2} & 2 & 0 & \frac{-5}{2} & 1 & 2 \\ & & 1 & \frac{1}{2} & -1 & 0 \\ \hline & 2 & 1 & -2 & 0 & 2 \end{array}$$

$y - r = y - \frac{1}{2}$,

so $r = \frac{1}{2}$.

The result is $2y^3 + y^2 - 2y + \dfrac{2}{y - \frac{1}{2}}$. Now simplify the fraction.

$$\frac{2}{y - \frac{1}{2}} = 2 \div \left(y - \frac{1}{2}\right)$$

$$= 2 \div \boxed{} = 2 \cdot \boxed{} \quad \text{or} \quad \boxed{}$$

The solution is $\boxed{}$.

Your Turn

a. Use long division to find $(x^2 + 5x + 6) \div (x + 3)$. $\boxed{}$

b. Use synthetic division to find $(16y^4 - 4y^2 + 2y + 8) \div (2y + 1)$.

$\boxed{}$

HOMEWORK ASSIGNMENT

Page(s):

Exercises:

Factoring Polynomials

WHAT YOU'LL LEARN

- Factor polynomials.
- Simplify polynomial quotients by factoring.

FOLDABLES

ORGANIZE IT

On the page for Lesson 5-4, write an ordered list describing what you look for as you factor a polynomial.

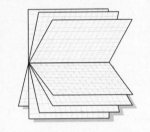

EXAMPLE GCF

1 Factor $10a^3b^2 + 15a^2b - 5ab^3$.

$10a^3b^2 + 15a^2b - 5ab^3$

$= (2 \cdot 5 \cdot a \cdot a \cdot a \cdot b \cdot b) + (3 \cdot 5 \cdot a \cdot a \cdot b) - (5 \cdot a \cdot b \cdot b \cdot b)$

$= (5ab \cdot \boxed{}) + (5ab \cdot \boxed{}) - (5ab \cdot \boxed{})$ The GCF is 5*ab*.

$= \boxed{}$ Distributive Property

EXAMPLE Grouping

2 Factor $x^3 + 5x^2 - 2x - 10$.

$x^3 + 5x^2 - 2x - 10$

$= \boxed{} + \boxed{}$ Group to find the GCF.

$= x^2 \boxed{} + (-2) \boxed{}$ Factor the GCF of each binomial.

$= \boxed{}$ Distributive Property

Your Turn Factor each polynomial.

a. $6x^4y^2 + 9x^2y^2 - 3xy^2$ $\boxed{}$

b. $x^3 - 3x^2 + 4x - 12$ $\boxed{}$

EXAMPLE Two or Three Terms

3 Factor each polynomial.

a. $3y^2 - 2y - 5$

The coefficient of the y terms must be 3 and -5 since $3(-5) = -15$ and $3 + (-5) = -2$. Rewrite the expression using $-5y$ and $3y$ in place of $-2y$ and factor by grouping.

$$3y^2 - 2y - 5 = 3y^2 \boxed{} - 5 \qquad \text{Replace } -2y.$$

$$= \boxed{} \qquad \begin{array}{l}\text{Associative} \\ \text{Property}\end{array}$$

$$= \boxed{} \qquad \text{Factor.}$$

$$= \boxed{} \qquad \begin{array}{l}\text{Distributive} \\ \text{Property}\end{array}$$

b. $x^3y^3 + 8$

$$x^3y^3 + 8 = (xy)^3 + 2^3 \qquad \begin{array}{l}\text{This is the sum} \\ \text{of two cubes.}\end{array}$$

$$= (xy + 2)\boxed{}$$

$$= \boxed{} \qquad \text{Simplify.}$$

Your Turn Factor each polynomial.

a. $2x^2 + x - 3$ \qquad $\boxed{}$

b. $3x^3 - 12x$ \qquad $\boxed{}$

c. $a^3b^3 - 27$ \qquad $\boxed{}$

EXAMPLE Quotient of Two Trinomials

④ Simplify $\dfrac{a^2 - a - 6}{a^2 + 7a + 10}$.

$$\frac{a^2 - a - 6}{a^2 + 7a + 10} = \frac{(a - 3)(a + 2)^{\scriptstyle 1}}{(a + 5)(a + 2)_{\scriptstyle 1}} \qquad \begin{array}{l}\text{Factor the numerator} \\ \text{and the denominator.}\end{array}$$

$$= \boxed{} \qquad \begin{array}{l}\text{Divide. Assume} \\ a \neq -5, -2.\end{array}$$

Your Turn Simplify $\dfrac{x^2 + 10x + 25}{x^2 + 3x - 10}$.

$$\boxed{}$$

HOMEWORK ASSIGNMENT

Page(s):

Exercises:

Roots of Real Numbers

WHAT YOU'LL LEARN

- Simplify radicals.
- Use a calculator to approximate radicals.

BUILD YOUR VOCABULARY (pages 96–97)

The inverse of raising a number to the *n*th power is finding the **nth root** of a number. When there is more than one real root, the [　　　　] root is called the **principal root**.

KEY CONCEPTS

Definition of Square Root
For any real numbers *a* and *b*, if $a^2 = b$, then *a* is a square root of *b*.

Definition of nth Root
For any real numbers *a* and *b*, and any positive integer *n*, if $a^n = b$, then *a* is an *n*th root of *b*.

EXAMPLE Find Roots

① Simplify

a. $\pm\sqrt{16x^6}$

$$\pm\sqrt{16x^6}$$

$$= \pm\sqrt{\boxed{}^2}$$

$$= \boxed{}$$

b. $-\sqrt{(q^3 + 5)^4}$

$$-\sqrt{(q^3 + 5)^4}$$

$$= -\sqrt{\left(\boxed{}\right)^2}$$

$$= \boxed{}$$

c. $\sqrt[5]{243a^{10}b^{15}}$

$$\sqrt[5]{243a^{10}b^{15}} = \sqrt[5]{\left(\boxed{}\right)^5}$$

$$= \boxed{}$$

d. $\sqrt{-4}$

$$\sqrt{-4} = \boxed{}$$

Since *n* is even and *b* is negative, $\sqrt{-4}$ is not a real number.

Your Turn Simplify.

a. $\pm\sqrt{9x^8}$

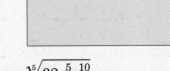

b. $-\sqrt{(a^3 + 2)^6}$

c. $\sqrt[5]{32x^5y^{10}}$

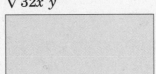

d. $\sqrt{-16}$

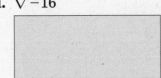

ORGANIZE IT

On the page for Lesson 5-5, explain how to simplify using absolute value. Include an example.

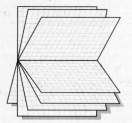

EXAMPLE Simplify Using Absolute Value

2 **Simplify.**

a. $\sqrt[6]{t^6}$

Note that t is a sixth root of t^6. The index is even, so the principal root is nonnegative. Since t could be negative, you must take the absolute value of t to identify the principal root.

$$\sqrt[6]{t^6} = \boxed{}$$

b. $\sqrt[5]{243(x + 2)^{15}}$

$$\sqrt[5]{243(x + 2)^{15}} = \sqrt[5]{\left[\boxed{}\right]^5}$$

Since the index is odd, you do not need absolute value.

$$\sqrt[5]{243(x + 2)^{15}} = \boxed{}$$

Your Turn Simplify.

a. $\sqrt[4]{x^4}$ $\boxed{}$ **b.** $\sqrt[3]{27(x + 2)^9}$ $\boxed{}$

EXAMPLE Approximate a Square Root

3 **PHYSICS** The time T in seconds that it takes a pendulum to make a complete swing back and forth is given by the formula $T = 2\pi\sqrt{\dfrac{L}{g}}$, where L is the length of the pendulum in feet and g is the acceleration due to gravity, 32 feet per second squared. Find the value of T for a 1.5-foot-long pendulum.

$T = 2\pi\sqrt{\dfrac{L}{g}}$ Original formula

$= \boxed{}$  $L = 1.5, g = 32$

$\approx \boxed{}$ seconds   Use a calculator.

HOMEWORK ASSIGNMENT

Page(s):

Exercises:

Your Turn Use the formula given in Example 3. Find the value of T for a 2-foot-long pendulum.

Radical Expressions

WHAT YOU'LL LEARN

- Simplify radical expressions.
- Add, subtract, multiply, and divide radical expressions.

KEY CONCEPTS

Product Property of Radicals For any real numbers a and b and any integer $n > 1$,

1. if n is even and a and b are both nonnegative, then $\sqrt[n]{ab} = \sqrt[n]{a} \cdot \sqrt[n]{b}$, and

2. if n is odd, then $\sqrt[n]{ab} = \sqrt[n]{a} \cdot \sqrt[n]{b}$.

Quotient Property of Radicals For any real numbers a and $b \neq 0$, and any integer $n > 1$, $\sqrt[n]{\dfrac{a}{b}} \cdot \dfrac{\sqrt[n]{a}}{\sqrt[n]{b}}$, if all roots are defined.

EXAMPLE Square Root of a Product

1 Simplify $\sqrt{25a^4b^9}$.

$$\sqrt{25a^4b^9} = \sqrt{\boxed{}(b^4)^2b}$$ Factor into squares where possible.

$$= \boxed{}\sqrt{(b^4)^2}\sqrt{b}$$ Product Property Radicals

$$= \boxed{}\,b^4\sqrt{b}$$ Simplify.

EXAMPLE Simplify Quotients

2 Simplify $\sqrt[3]{\dfrac{2}{9x}}$.

$$\sqrt[3]{\dfrac{2}{9x}} = \dfrac{\sqrt[3]{2}}{\sqrt[3]{9x}}$$ Quotient Property

$$= \dfrac{\sqrt[3]{2}}{\sqrt[3]{9x}} \cdot \dfrac{\sqrt[3]{3x^2}}{\sqrt[3]{3x^2}}$$ Rationalize the denominator.

$$= \dfrac{\sqrt[3]{2 \cdot 3x^2}}{\sqrt[3]{9x \cdot 3x^2}}$$ Product Property

$$= \boxed{}$$ Multiply.

$$= \boxed{}$$ $\sqrt[3]{27x^3} = 3x$

Your Turn Simplify each expression.

a. $\sqrt{16x^4y^{11}}$

b. $\sqrt{\dfrac{x^3}{y^7}}$

c. $\sqrt[3]{\dfrac{2}{3a}}$

BUILD YOUR VOCABULARY (page 96)

Two radical expressions are called **like radical expressions** if both the indices and radicands are alike.

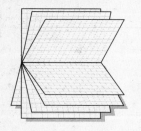
EXAMPLE Add and Subtract Radicals

3 Simplify $3\sqrt{45} - 5\sqrt{80} + 4\sqrt{20}$.

$3\sqrt{45} - 5\sqrt{80} + 4\sqrt{20}$

$= 3\sqrt{3^2 \cdot 5} - 5\sqrt{4^2 \cdot 5} + 4\sqrt{2^2 \cdot 5}$ Factor.

$= 3\sqrt{3^2} \cdot \sqrt{5} - 5\sqrt{4^2} \cdot \sqrt{5} + 4\sqrt{2^2} \cdot \sqrt{5}$ Product Property

$= 3 \cdot \boxed{}\sqrt{5} - 5 \cdot 4\sqrt{5} + 4 \cdot \boxed{}\sqrt{5}$

$= \boxed{}$ Multiply.

$= \boxed{}$ Combine like radicals.

EXAMPLE Multiply Radicals

4 Simplify $(2\sqrt{3} + 3\sqrt{5})(3 - \sqrt{3})$.

$(2\sqrt{3} + 3\sqrt{5})(3 - \sqrt{3})$

    F   O   I   L

$= 2\sqrt{3} \cdot 3 - \boxed{} + \boxed{} - 3\sqrt{5} \cdot \sqrt{3}$

$= 6\sqrt{3} - \boxed{} + \boxed{} - 3\sqrt{15}$

$= 6\sqrt{3} - \boxed{} - 3\sqrt{15}$

Your Turn 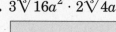 Simplify each expression.

a. $3\sqrt[3]{16a^2} \cdot 2\sqrt[3]{4a}$

b. $3\sqrt{75} - 2\sqrt{48} + \sqrt{3}$

c. $(2\sqrt{5} + 4\sqrt{6})(5 - \sqrt{7})$

Rational Exponents

WHAT YOU'LL LEARN

- Write expressions with rational exponents in radical form, and vice versa.

- Simplify expressions in exponential or radical form.

KEY CONCEPTS

$b^{\frac{1}{n}}$ For any real number b and for any positive integer n, $b^{\frac{1}{n}} = \sqrt[n]{b}$, except when $b < 0$ and n is even.

Rational Exponents For any nonzero real number b, and any integers m and n, with $n > 1$, $b^{\frac{m}{n}} = \sqrt[n]{b^m} = \left(\sqrt[n]{b}\right)^m$, except when $b < 0$ and n is even.

EXAMPLE Radical Form

1 Write each expression in radical form.

a. $a^{\frac{1}{6}}$

$a^{\frac{1}{6}} = $

b. $m^{\frac{1}{2}}$

$m^{\frac{1}{2}} = $

Your Turn Write each expression in radical form.

a. $a^{\frac{1}{3}}$

b. $x^{\frac{1}{5}}$

EXAMPLE Exponential Form

2 Write each radical using rational exponents.

a. $\sqrt[5]{b}$

$\sqrt[5]{b} = $

b. $\sqrt{w}$

$\sqrt{w} = $

Your Turn Write each radical using rational exponents.

a. $\sqrt[4]{x}$

b. $\sqrt[3]{y}$

EXAMPLE Evaluate Expressions with Rational Exponents

3 Evaluate each expression.

a. $49^{-\frac{1}{2}}$

$49^{-\frac{1}{2}} = \dfrac{1}{49^{\frac{1}{2}}}$

$b^{-n} = \dfrac{1}{b^n}$

$= \dfrac{1}{\boxed{}}$

$49^{\frac{1}{2}} = $

$= \boxed{}$

Simplify.

b. $32^{\frac{2}{5}}$

$32^{\frac{2}{5}} = (2^5)^{\frac{2}{5}}$

$= \boxed{}$

$= \boxed{}$

$= \boxed{}$

Your Turn Evaluate each expression.

a. $25^{-\frac{1}{2}}$

b. $16^{\frac{3}{4}}$

EXAMPLE Simplify Expressions with Rational Exponents

4 Simplify each expression.

a. $y^{\frac{1}{7}} \cdot y^{\frac{4}{7}}$

$y^{\frac{1}{7}} \cdot y^{\frac{4}{7}} = \boxed{}$ Multiply powers.

$= \boxed{}$ Add exponents.

b. $x^{-\frac{2}{3}}$

$x^{-\frac{2}{3}} = \dfrac{1}{x^{\frac{2}{3}}}$ $b^{-n} = \dfrac{1}{b^n}$

$= \dfrac{1}{x^{\frac{2}{3}}} \cdot \boxed{}$ Multiply by $\boxed{}$.

$= \boxed{}$ or $\boxed{}$ $x^{\frac{2}{3}} \cdot x^{\frac{1}{3}} = x^{\frac{2}{3} + \frac{1}{3}}$

Your Turn Simplify each expression.

a. $x^{\frac{1}{5}} \cdot x^{\frac{2}{5}}$

b. $y^{-\frac{3}{4}}$

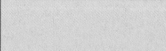

EXAMPLE **Simplify Radical Expressions**

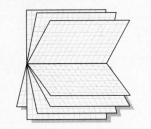

FOLDABLES

ORGANIZE IT

On the page for Lesson 5-7, write four conditions that must be met in order for an expression with rational exponents to be in simplest form.

5 **Simplify each expression.**

a. $\dfrac{\sqrt[6]{16}}{\sqrt[3]{2}}$

$\dfrac{\sqrt[6]{16}}{\sqrt[3]{2}} = \dfrac{16^{\frac{1}{6}}}{2^{\frac{1}{3}}}$ Rational exponents

$= \dfrac{\boxed{}}{2^{\frac{1}{3}}}$ $16 = 2^4$

$= \dfrac{\boxed{}}{2^{\frac{1}{3}}}$ Power of a Power

$= \boxed{}$ Quotient of Powers

$= \boxed{}$ or $\boxed{}$ Simplify.

b. $\sqrt[6]{4x^4}$

$\sqrt[6]{4x^4} = (4x^4)^{\frac{1}{6}}$ Rational exponents

$= (2^2 \boxed{})^{\frac{1}{6}}$ $2^2 = 4$

$= 2^{2\left(\frac{1}{6}\right)} \cdot \boxed{}$ Power of a Power

$= \boxed{}$ Multiply.

$= \boxed{}$ Simplify.

Your Turn **Simplify each expression.**

a. $\dfrac{\sqrt[4]{4}}{\sqrt{2}}$ **b.** $\sqrt[3]{16x^2}$

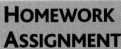

HOMEWORK ASSIGNMENT

Page(s):

Exercises:

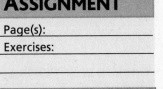

Radical Equations and Inequalities

BUILD YOUR VOCABULARY (pages 96–97)

Equations with radicals that have variables in the radicands are called **radical equations**.

When you solve a radical equation and obtain a number that does not satisfy the original equation, the number is called an **extraneous solution**.

A **radical inequality** is an inequality that has a variable in a radicand.

EXAMPLE Solve a Radical Equation

1 Solve $\sqrt{y-2} - 1 = 5$.

$\sqrt{y-2} - 1 = 5$			Original equation
	=		Add to isolate the radical.
	=		Square each side.
	=		Find the squares.
$y = 38$			Add 2 to each side.

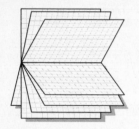

EXAMPLE Extraneous Solution

2 Solve $\sqrt{x-12} = 2 - \sqrt{x}$.

$\sqrt{x-12} = 2 - \sqrt{x}$		Original equation
$(\sqrt{x-12})^2 = (2-\sqrt{x})^2$		Square each side.
$x - 12 =$		Find the squares.
$-16 = -4\sqrt{x}$		Isolate the radical.
$ = $		Divide each side by -4.
$ = $		Square each side.
$16 = x$		Evaluate the squares.

Since $\sqrt{16-12} \neq$ [   ], the solution does not check

and there is [   ].

© Glencoe/McGraw-Hill

Your Turn Solve.

a. $\sqrt{x-3}-2=6$

b. $\sqrt{x+5}=-1-\sqrt{x}$

c. $(2y+1)^{\frac{1}{3}}-3=0$

EXAMPLE Radical Inequality

3 Solve $\sqrt{3x-6}+4\le 7$.

Find values of x for which the left side is defined.

$3x-6\ge 0$ Radicand must be positive or 0.

$\quad 3x\ge 6$

 $\ge$

Now solve $\sqrt{3x-6}+4\le 7$.

$\sqrt{3x-6}+4\le 7$ Original equation

$\qquad \sqrt{3x-6}\le 3$ Isolate the radical.

$\qquad \boxed{}\le\boxed{}$ Eliminate the radical.

$\qquad\qquad 3x\le 15$ Add 6 to each side.

$\qquad\qquad\quad x\le 5$ Divide each side by 3.

The solution is 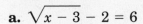.

Your Turn Solve.

a. $\sqrt{x-3}-2=6$

b. $\sqrt{x+5}=-1-\sqrt{x}$

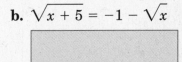

c. $(2y+1)^{\frac{1}{3}}-3=0$

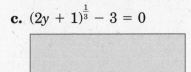

d. $\sqrt{2x+5}-2\le 9$

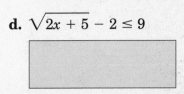

HOMEWORK ASSIGNMENT

Page(s): _____

Exercises: _____

Complex Numbers

WHAT YOU'LL LEARN

- Add and subtract complex numbers.

- Multiply and divide complex numbers.

BUILD YOUR VOCABULARY (pages 96–97)

i is called the **imaginary unit** ($i = \sqrt{-1}$).

Pure imaginary numbers are square roots of [   ] real numbers.

A **complex number**, such as the expression $5 + 2i$, is a complex number since it has a real number (5) and a pure imaginary number ($2i$).

Two complex numbers of the form $a + bi$ and [   ] are called **complex conjugates**.

EXAMPLE Square Roots of Negative Numbers

1 Simplify $\sqrt{-32y^3}$.

$$\sqrt{-32y^3} = \sqrt{-1 \cdot 4^2 \cdot 2 \cdot y^2 \cdot y}$$

$$= \sqrt{-1} \cdot \sqrt{4^2} \cdot \sqrt{2} \cdot \sqrt{y^2} \cdot \sqrt{y}$$

$$= \boxed{} \cdot 4|y|\sqrt{2} \cdot \boxed{}$$

$$= \boxed{}$$

EXAMPLE Multiply Pure Imaginary Numbers

2 Simplify.

a. $-3i \cdot 2i$

$$-3i \cdot 2i = 6i^2$$

$$= -6\left(\boxed{}\right) \qquad\qquad i^2 = \boxed{}$$

$$= \boxed{}$$

REMEMBER IT

You can write i to the left or right of the radical symbol. However, i is usually written to the left so it is clear that it is not under the radical.

b. $\sqrt{-12} \cdot \sqrt{-2}$

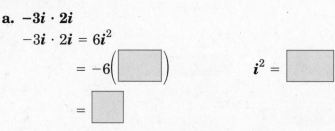

$$\sqrt{-12} \cdot \sqrt{-2} = \boxed{} \cdot \boxed{}$$

$$= i^2\sqrt{24}$$

$$= \boxed{} \quad\text{or}\quad \boxed{}$$

EXAMPLE Simplify a Power of *i*

3 Simplify i^{35}.

$i^{35} = i \cdot i^{34}$ Multiplying powers

$= i \cdot \boxed{}^{17}$ Power of a Power

$= i \cdot \boxed{}^{17}$ $i^2 = \boxed{}$

$= i \cdot \boxed{}$ or $-i$ $(-1)^{17} = \boxed{}$

Your Turn Simplify.

a. $\sqrt{-50x^5}$ **b.** $3i \cdot 5i$

c. $\sqrt{-2} \cdot \sqrt{-6}$ **d.** i^{37}

KEY CONCEPT

Complex Numbers A complex number is any number that can be written in the form $a + bi$, where a and b are real numbers and i is the imaginary unit. a is called the real part, and b is called the imaginary part.

EXAMPLE Add and Subtract Complex Numbers

4 Simplify $(4 - 6i) - (3 - 7i)$.

$(4 - 6i) - (3 - 7i) = \boxed{}$

$= \boxed{}$

Your Turn Simplify $(3 + 2i) - (-2 + 5i)$.

EXAMPLE Multiply Complex Numbers

5 ELECTRICITY In an AC circuit, the voltage E, current I, and impedance Z are related to the formula $E = I \cdot Z$. Find the voltage in a circuit with current $1 + 4j$ amps and impedance $3 - 6j$ ohms.

$E = I \cdot Z$	Electricity formula
$= (1 + 4j)(3 - 6j)$	$I = 1 + 4j, Z = 3 - 6j$
$= 1(3) + 1(-6j) + 4j(3) + 4j(-6j)$	FOIL
$= 3 - \boxed{} + 12j - \boxed{}$	Multiply
$= \boxed{}$	$j^2 = -1$
$= \boxed{}$ volts	Add.

Your Turn Refer to Example 5. Find the voltage in a circuit with current $1 - 3j$ amps and impedance $3 + 2j$ ohms.

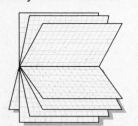
EXAMPLE Divide Complex Numbers

6 Simplify $\dfrac{5i}{3 + 2i}$.

$\dfrac{5i}{3 + 2i} = \dfrac{5i}{3 + 2i} \cdot \boxed{}$	$3 - 2i$ and $3 + 2i$ are conjugates.
$= \dfrac{15i - 10i^2}{9 - 4i^2}$	Multiply
$= \dfrac{15i + 10}{\boxed{}}$	$i^2 = -1$
$= \boxed{} + \boxed{}$	Standard form

Your Turn Simpify $\dfrac{3i}{1 + i}$.

HOMEWORK ASSIGNMENT

Page(s): _____

Exercises: _____

_____

BRINGING IT ALL TOGETHER

FOLDABLES™	VOCABULARY PUZZLEMAKER	BUILD YOUR VOCABULARY
Use your **Chapter 5 Foldable** to help you study for your chapter test.	To make a crossword puzzle, word search, or jumble puzzle of the vocabulary words in Chapter 5, go to: www.glencoe.com/sec/math/ t_resources/free/index.php	You can use your completed **Vocabulary Builder** (pages 96–97) to help you solve the puzzle.

5-1
Monomials

Simplify. Assume that no variable equals 0.

1. $(3n^4y^3)(-2ny^{-5})$

2. $\dfrac{12(x^2y)^3}{4(xy^0)^2}$

5-2
Polynomials

Determine whether each expression is a polynomial. If the expression is a polynomial, classify it by the number of terms and state the degree of the polynomial.

3. $\sqrt{3x}$

4. $4r^4 - 2r + 1$

5. $2ab + 4ab^2 - 6ab^3$

6. $5x + 4y$

Simplify.

7. $(3a - 6) - (2a - 1)$

8. $(2x - 5)(3x + 5)$

5-3
Dividing Polynomials

Simplify.

9. $(c^3 + c^2 - 14c - 24) \div (c - 4)$

10. $\dfrac{n^2 + 3n - 2}{n + 2}$

5-4
Factoring Polynomials

Factor completely. If the polynomial is not factorable, write *prime*.

11. $3w^2 - 48$

12. $a^3 + 5a - 3a^2 - 15$

13. Simplify $\dfrac{x^2 + 7x + 10}{x^2 - 4}$. Assume that the denominator is not equal to 0.

5-5
Roots of Real Numbers

14. Use a calculator to approximate $\sqrt[3]{-280}$ to three decimal places.

Simplify.

15. $\sqrt[3]{-64}$

16. $\sqrt[8]{x^{16}}$

17. $\sqrt{100a^{12}}$

18. $\sqrt[3]{8y^3}$

19. $\sqrt{49x^6y^8}$

20. $\sqrt[3]{125c^6d^{15}}$

5-6
Radical Expressions

Simplify.

21. $\sqrt{\dfrac{5}{6x}}$

22. $2\sqrt{45} - 7\sqrt{8} + \sqrt{80}$

5-7

Rational Exponents

Evaluate.

23. $(-32)^{\frac{1}{5}}$

24. $25^{-\frac{3}{2}}$

25. $\left(\frac{1}{64}\right)^{-\frac{1}{3}}$

Simplify.

26. $x^{\frac{1}{4}} \cdot x^{\frac{5}{4}}$

27. $\left(y^{\frac{-5}{6}}\right)^{\frac{-1}{5}}$

28. $\sqrt[10]{36a^2b^{10}}$

5-8

Radical Equations and Inequalities

Solve each equation or inequality.

29. $\sqrt[3]{5u-2} = -3$

30. $\sqrt{4z-3} = \sqrt{9z+2}$

31. $3 + \sqrt{2x-1} \le 6$

32. $\sqrt{5x+4} + 9 > 13$

5-9

Complex Numbers

Simplify.

33. $\sqrt{-2} \cdot \sqrt{-10}$

34. $(3 - 8i) - (5 + 2i)$

35. $(4 - i)(5 + 2i)$

36. $\frac{3 - i}{2 + i}$

37. Solve $5x^2 + 60 = 0$.

ARE YOU READY FOR THE CHAPTER TEST?

Check the one that applies. Suggestions to help you study are given with each item.

☐ **I completed the review of all or most lessons without using my notes or asking for help.**

- You are probably ready for the Chapter Test.

- You may want to take the Chapter 5 Practice Test on page 281 of your textbook as a final check.

☐ **I used my Foldable or Study Notebook to complete the review of all or most lessons.**

- You should complete the Chapter 5 Study Guide and Review on pages 276–280 of your textbook.

- If you are unsure of any concepts or skills, refer back to the specific lesson(s).

- You may also want to take the Chapter 5 Practice Test on page 281.

☐ **I asked for help from someone else to complete the review of all or most lessons.**

- You should review the examples and concepts in your Study Notebook and Chapter 5 Foldable.

- Then complete the Chapter 5 Study Guide and Review on pages 276–280 of your textbook.

- If you are unsure of any concepts or skills, refer back to the specific lesson(s).

- You may also want to take the Chapter 1 Practice Test on page 281.

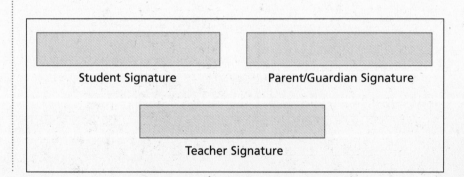

Student Signature   Parent/Guardian Signature

Teacher Signature

Quadratic Functions and Inequalities

 Use the instructions below to make a Foldable to help you organize your notes as you study the chapter. You will see Foldable reminders in the margin of this Interactive Study Notebook to help you in taking notes.

Begin with one sheet of 11"   17" paper.

STEP 1 **Fold and Cut**
Fold in half lengthwise. Then fold in fourths crosswise. Cut along the middle fold from the edge to the last crease as shown.

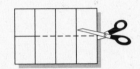

STEP 2 **Refold and Label**
Refold along lengthwise fold and staple uncut section at top. Label the section with a lesson number and close to form a booklet.

 NOTE-TAKING TIP: When you take notes, think about the order in which the concepts are being presented. Write why you think the concepts were presented in this sequence.

BUILD YOUR VOCABULARY

This is an alphabetical list of new vocabulary terms you will learn in Chapter 6. As you complete the study notes for the chapter, you will see Build Your Vocabulary reminders to complete each term's definition or description on these pages. Remember to add the textbook page number in the second column for reference when you study.

Vocabulary Term	Found on Page	Definition	Description or Example
axis of symmetry			
completing the square			
constant term			
discriminant [dihs-KRIH-muh-nuhnt]			
linear term			
maximum value			
minimum value			
parabola [puh-RA-buh-luh]			
quadratic equation [kwah-DRA-tihk]			

© Glencoe/McGraw-Hill

Vocabulary Term	Found on Page	Definition	Description or Example
Quadratic Formula			
quadratic function			
quadratic inequality			
quadratic term			
roots			
Square Root Property			
vertex			
vertex form			
Zero Product Property			
zeros			

Graphing Quadratic Functions

EXAMPLE Graph a Quadratic Function

WHAT YOU'LL LEARN

- Graph quadratic functions.

- Find and interpret the maximum and minimum values of a quadratic function.

KEY CONCEPT

Graph of a Quadratic Function Consider the graph of
$y = ax^2 + bx + c$,
where $a \neq 0$.

- The y-intercept is $a(0)^2 + b(0) + c$ or c.

- The equation of the axis of symmetry is $x = -\dfrac{b}{2a}$.

- The x-coordinate of the vertex is $-\dfrac{b}{2a}$.

1 Graph $f(x) = x^2 + 3x - 1$ by making a table of values.

First, choose integer values for x. Then, evaluate the function for each x value. Graph the function.

x	$x^2 + 3x - 1$	$f(x)$	(x, y)
-3	$(-3)^2 + 3(-3) - 1$		
-2	$(-2)^2 + 3(-2) - 1$		
-1	$(-1)^2 + 3(-1) - 1$		
0	$(0)^2 + 3(0) - 1$		
1	$(1)^2 + 3(1) - 1$		

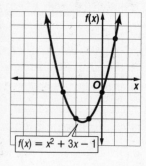

$f(x) = x^2 + 3x - 1$

Your Turn Graph $f(x) = 2x^2 + 3x + 2$.

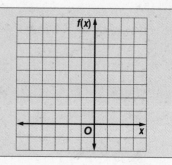

EXAMPLE Axis of Symmetry, y-intercept, and Vertex

2 Consider the quadratic function $f(x) = 2 - 4x + x^2$.

a. Find the y-intercept, the equation of the axis of symmetry, and the x-coordinate of the vertex.

Begin by rearranging the terms of the function. Then identify a, b, and c.

$$f(x) = ax^2 + bx + c$$

$$f(x) = 2 - 4x + x^2 \longrightarrow f(x) = 1x^2 - 4x + 2$$

So, $a = \boxed{}$, $b = \boxed{}$, and $c = \boxed{}$.

The y-intercept is 2. You can find the equation of the axis of symmetry by using a and b.

$x = -\dfrac{b}{2a}$ Equation of the axis of symmetry

$x =$ [   ] $a = 1$, $b = -4$

$x =$ [ ]   Simplify.

The y-intercept is [ ]. The equation of the axis of

symmetry is $x =$ [ ]. Therefore, the x-coordinate of the

vertex is [ ].

b. Make a table of values that includes the vertex.

Choose some values for x that are less than 2 and some that are greater than 2.

x	$x^2 - 4x + 2$	$f(x)$	$(x, f(x))$
0	$0^2 - 4(0) + 2$		
1	$1^2 - 4(1) + 2$		
2	$2^2 - 4(2) + 2$		
3	$3^2 - 4(3) + 2$		
4	$4^2 - 4(4) + 2$		

c. Use this information to graph the function.

Graph the vertex and the y-intercept.

Then graph the points from your table connecting them with a smooth curve.

As a check, draw the axis of symmetry, $x = 2$, as a dashed line.

The graph of the function should be symmetric about this line.

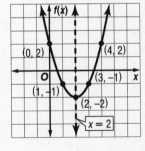

Your Turn Consider the quadratic function $f(x) = 3 - 6x + x^2$.

a. Find the y-intercept, the equation of the axis of symmetry and the x-coordinate of the vertex.

WRITE IT

Why is it helpful to know the axis of symmetry when graphing a quadratic function?

b. Make a table of values that includes the vertex.

c. Use this information to graph the function.

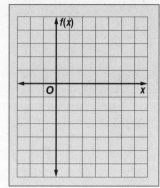

EXAMPLE Maximum or Minimum Value

3 Consider the function $f(x) = -x^2 + 2x + 3$.

a. Determine whether the function has a maximum or a minimum value.

For this function, $a = \boxed{}$, $b = \boxed{}$, and $c = \boxed{}$.

Since $a \boxed{}$ 0, the graph opens $\boxed{}$ and the

function has a $\boxed{}$.

b. State the maximum or minimum value of the function.

The maximum value of this function is the y-coordinate of the vertex.

The x-coordinate of the vertex is $\boxed{}$ or $\boxed{}$.

Find the y-coordinate of the vertex by evaluating the function for $x = 1$.

$f(x) = x^2 + 2x + 3$ Original function

$f(1) = \boxed{}$ or $\boxed{}$ $x = 1$

The maximum value of the function is $\boxed{}$.

Your Turn Consider the function $f(x) = x^2 + 4x - 1$.

a. Determine whether the function has a maximum or a minimum value. $\boxed{}$

b. State the maximum or minimum value of the function.

$\boxed{}$

Solving Quadratic Equations by Graphing

WHAT YOU'LL LEARN

- Solve quadratic equations by graphing.

- Estimate solutions of quadratic equations by graphing.

BUILD YOUR VOCABULARY (pages 124–125)

A **quadratic equation** can be written in the form $ax^2 + bx + c = 0$, where $a \neq 0$.

The [   ] of a quadratic equation are called the **roots** of the equation. One method for finding the roots of a quadratic [   ] is to find the **zeros** of the related quadratic [   ].

EXAMPLE Two Real Solutions

1 **Solve $x^2 - 3x - 4 = 0$ by graphing.**
Graph the related quadratic function $f(x) = x^2 - 3x - 4$. The equation of the axis of symmetry is $x = -\dfrac{-3}{2(1)}$ or $\dfrac{3}{2}$. Make a table using x-values around $x = \dfrac{3}{2}$. Then graph each point.

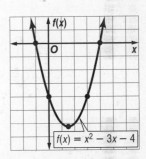

$f(x) = x^2 - 3x - 4$

x	−1	0	1	2	3	4
f(x)						

From the table and the graph, we can see that the zeros of the function are −1 and 4. The solutions of the equation are

[   ] and [   ].

Your Turn Solve $x^2 + 2x - 3 = 0$ by graphing.

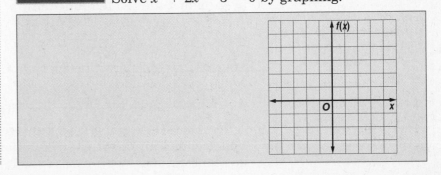

EXAMPLE One Real Solution

2 Solve $x^2 - 4x = -4$ by graphing.

Write the equation in $ax^2 + bx + c = 0$ form.

$x^2 - 4x = -4 \longrightarrow$ [   ] $= 0$ Add 4 to each side.

Graph the related quadratic function $f(x) = x^2 - 4x + 4$.

x	0	1	2	3	4
$f(x)$					

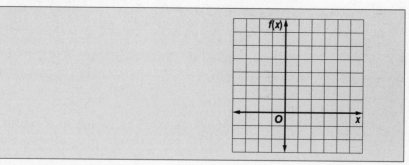

$f(x) = x^2 - 4x + 4$

Notice that the graph has only one x-intercept, 2.

Your Turn Solve $x^2 - 6x = -9$ by graphing.

EXAMPLE No Real Solution

3 NUMBER THEORY Find two real numbers whose sum is 4 and whose product is 5 or show that no such numbers exist.

Let x = one of the numbers. Then $4 - x$ = the other number.

Since the product is 5, you know that $x(4 - x) = 5$ or $-x^2 + 4x - 5 = 0$.

You can solve $x^2 - 4x + 5 = 0$ by graphing the related function $f(x) = x^2 - 4x + 5$.

x	0	1	2	3	4
$f(x)$					

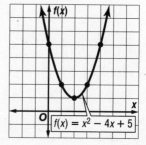

$f(x) = x^2 - 4x + 5$

Notice that the graph has no x-intercepts. This means that the original equation has no real solution.

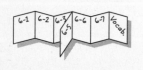

© Glencoe/McGraw-Hill

Your Turn Find two real numbers whose sum is 7 and whose product is 14 or show that no such numbers exit.

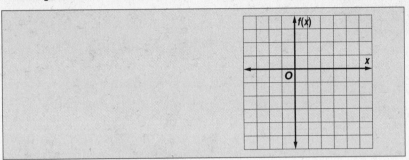

EXAMPLE Estimate Roots

④ Solve $x^2 - 6x + 3 = 0$ by graphing. If exact roots cannot be found, state the consecutive integers between which the roots are located.

The equation of the axis of symmetry of the related function is

$x =$ _____ .

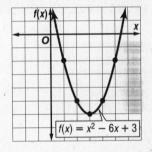

$f(x) = x^2 - 6x + 3$

x	0	1	2	3	4	5	6
f(x)							

The x-intercepts of the graph are between 0 and 1 and between 5 and 6.

Your Turn Solve $x^2 - 4x + 2 = 0$ by graphing. If exact roots cannot be found, state the consecutive integers between which the roots are located.

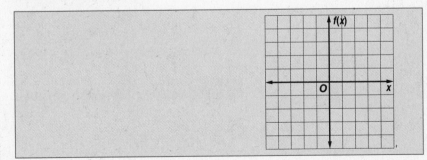

HOMEWORK ASSIGNMENT

Page(s):
Exercises:

Solving Quadratic Equations by Factoring

WHAT YOU'LL LEARN

- Solve quadratic equations by factoring.

- Write a quadratic equation with given roots.

KEY CONCEPT

Zero Product Property
For any real numbers a and b, if $ab = 0$, then either $a = 0$, $b = 0$, or both a and b equal zero.

EXAMPLE Two Roots

① Solve each equation by factoring.

a. $x^2 = -4x$

$x^2 = -4x$	Original equation
$\boxed{} = 0$	Add $4x$ to each side.
$\boxed{} = 0$	Factor the binomial.
$\boxed{} = 0$ or $\boxed{} = 0$	Zero Product Property
$x = \boxed{}$	Solve the second equation.

The solution set is $\boxed{}$. Check the solution.

b. $3x^2 = 5x + 2$

$3x^2 = 5x + 2$	Original equation
$\boxed{} = 0$	Subtract $5x$ and 2 from each side.
$\boxed{} = 0$	Factor the trinomial.
$\boxed{} = 0$ or $\boxed{} = 0$	Zero Product Property
$\boxed{} = \boxed{}$ $x = \boxed{}$	Solve each equation.
$x = \boxed{}$	

The solution set is $\boxed{}$.

Your Turn Solve each equation by factoring.

a. $x^2 = 3x$

b. $6x^2 + 11x = -4$

EXAMPLE Double Root

2 Solve $x^2 - 6x = -9$ by factoring.

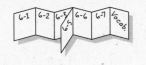
$$x^2 - 6x = -9 \quad \text{Original equation}$$

$$\boxed{} = 0 \quad \text{Add 9 to each side.}$$

$$\boxed{} = 0 \quad \text{Factor.}$$

$$\boxed{} = 0 \text{ or } \boxed{} = 0 \quad \text{Zero Product Property}$$

$$x = 3 \qquad\qquad x = 3 \quad \text{Solve each equation.}$$

Your Turn Solve $x^2 + 10x = -25$ by factoring.

EXAMPLE Write an Equation Given Roots

3 Write a quadratic equation with $-\frac{2}{3}$ and 6 as its roots. Write the equation in the form $ax^2 + bx + c = 0$, where a, b, and c are integers.

$$(x - p)(x - q) = 0 \quad \text{Write the pattern.}$$

$$\left[x - \left(-\frac{2}{3}\right)\right](x - 6) = 0 \quad \text{Replace } p \text{ with } -\frac{2}{3} \text{ and } q \text{ with 6.}$$

$$\boxed{} = 0 \quad \text{Simplify.}$$

$$\boxed{} = 0 \quad \text{Use FOIL.}$$

$$\boxed{} = 0 \quad \text{Multiply each side by 3 so that } b \text{ is an integer.}$$

HOMEWORK ASSIGNMENT
Page(s):
Exercises:

Your Turn Write a quadratic equation with $-\frac{3}{4}$ and 5 as its roots. Write the equation in the form $ax^2 + bx + c = 0$, where a, b, and c are integers.

Completing the Square

WHAT YOU'LL LEARN

- Solve quadratic equations by using the Square Root Property.

- Solve quadratic equations by completing the square.

KEY CONCEPT

Square Root Property

For any real number n, if $x^2 = n$, then $x = \pm\sqrt{n}$.

EXAMPLE Equation with Rational Roots

1 Solve $x^2 + 14x + 49 = 64$ by using the Square Root Property.

$$x^2 + 14x + 49 = 64 \qquad \text{Original equation}$$

$$\boxed{} = 64 \qquad \text{Factor the trinomial.}$$

$$\boxed{} = \pm\boxed{} \qquad \text{Square Root Property}$$

$$\boxed{} = \boxed{} \qquad \sqrt{64} = 8$$

$$x = -7 \pm 8 \qquad \text{Subtract 7 from each side.}$$

$$x = -7 + 8 \text{ or } x = -7 - 8 \qquad \text{Write as two equations.}$$

$$x = \boxed{} \qquad x = \boxed{} \qquad \text{Solve each equation.}$$

EXAMPLE Equation with Irrational Roots

2 Solve $x^2 - 10x + 25 = 12$ by using the Square Root Property.

$$x^2 - 10x + 25 = 12 \qquad \text{Original equation}$$

$$(x - 5)^2 = 12 \qquad \text{Factor the trinomial.}$$

$$x - 5 = \boxed{} \qquad \text{Square Root Property}$$

$$x = \boxed{} \qquad \begin{array}{l}\text{Add 5 to each side.} \\ \sqrt{12} = 2\sqrt{3}\end{array}$$

$$x = 5 + \boxed{} \text{ or } x = 5 - \boxed{} \qquad \text{Write as two equations.}$$

$$x \approx \boxed{} \qquad x \approx \boxed{} \qquad \text{Use a calculator.}$$

Your Turn Solve by using the Square Root Property.

a. $x^2 - 16x + 64 = 25$

b. $x^2 - 4x + 4 = 8$

EXAMPLE Complete the Square

③ Find the value of c that makes $x^2 + 16x + c$ a perfect square. Then write the trinomial as a perfect square.

STEP 1 Find one half of 16.

STEP 2 Square the result of Step 1.

STEP 3 Add the result of Step 2 to $x^2 + 16x$.

The trinomial [ ] can be written as

[ ].

Your Turn Find the value of c that makes $x^2 + 6x + c$ a perfect square. Then write the trinomial as a perfect square.

KEY CONCEPT

Completing the Square
To complete the square for any quadratic expression of the form $x^2 + bx$, follow the steps below.

Step 1 Find one half of b, the coefficient of x.

Step 2 Square the result in Step 1.

Step 3 Add the result of Step 2 to $x^2 + bx$.

EXAMPLE Solve an Equation by Completing the Square

④ Solve $x^2 + 4x - 12 = 0$ by completing the square.

$x^2 + 4x - 12 = 0$ — Notice that $x^2 + 4x - 12$ is not a perfect square.

$x^2 + 4x = $ [ ] — Rewrite so the left side is of the form $x^2 + bx$.

$x^2 + 4x + 4 = 12 + 4$ — Add [ ] to each side.

[ ] $= 16$ — Write the left side as a perfect square by factoring.

$x + 2 = $ [ ] — Square Root Property

$x = $ [ ] — Subtract 2 from each side.

$x = $ [ ] or $x = $ [ ] — Write as two equations.

$x = $ [ ] $x = $ [ ] — Solve each equation.

REMEMBER IT

Be sure to add the same constant to both sides of the equation when solving equations by completing the square.

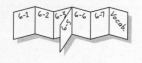

ORGANIZE IT

On the page for Lesson 6-4, list the steps you would use to solve $w^2 - 8w - 9 = 0$ by completing the square.

EXAMPLE Equation with $a \neq 1$

5 Solve $3x^2 - 2x - 1 = 0$ by completing the square.

$$3x^2 - 2x - 1 = 0$$

Notice that $3x^2 - 2x - 1$ is not a perfect square.

$$x^2 - \frac{2}{3}x - \frac{1}{3} = 0$$

Divide by the coefficient of the quadratic term, 3.

$$x^2 - \frac{2}{3}x = \frac{1}{3}$$

Add $\frac{1}{3}$ to each side.

$$x^2 - \frac{2}{3}x + \boxed{} = \frac{1}{3} + \boxed{}$$

Since $\left(\dfrac{-2}{3} \cdot \dfrac{1}{2}\right)^2 = \boxed{}$,

add $\boxed{}$ to each side.

Write the left side as a perfect square by factoring. Simplify the right side.

$$x - \frac{1}{3} = \pm\frac{2}{3}$$

Square Root Property

$$x = \boxed{} \quad \text{or } x = \boxed{}$$

Write as two equations.

$$x = \boxed{} \qquad x = \boxed{}$$

Solve each equation.

Your Turn Solve each equation by completing the square.

a. $x^2 + 5x - 6 = 0$

b. $2x^2 + 11x + 15 = 0$

c. $x^2 + 4x + 5 = 0$

HOMEWORK ASSIGNMENT

Page(s):

Exercises:

The Quadratic Formula and the Discriminant

WHAT YOU'LL LEARN

- Solve quadratic equations by using the Quadratic Formula.

- Use the discriminant to determine the number and types of roots of a quadratic equation.

KEY CONCEPT

Quadratic Formula
The solutions of a quadratic equation of the form
$ax^2 + bx + c = 0$, where $a \neq 0$, are given by the following formula.

$x = \dfrac{-b \pm \sqrt{b^2 - 4ac}}{2a}$

EXAMPLE Two Rational Roots

1 Solve $x^2 - 8x = 33$ by using the Quadratic Formula.

First, write the equation in the form $ax^2 + bx + c = 0$ and identify a, b, and c.

$$ax^2 \quad + \quad bx \quad + \quad c = 0$$

$$x^2 - 8x = 33 \longrightarrow \boxed{} - \boxed{} - \boxed{} = 0$$

Then, substitute these values into the Quadratic Formula.

$x = \dfrac{-b \pm \sqrt{b^2 - 4ac}}{2a}$ Quadratic Formula

$x = \dfrac{-(-8) \pm \sqrt{(-8)^2 - 4(1)(-33)}}{2(1)}$ Replace a with 1, b with -8, and c with -33.

$x = \dfrac{8 \pm \sqrt{\boxed{} + \boxed{}}}{2}$ Simplify.

$x = \dfrac{8 \pm \sqrt{\boxed{}}}{2}$ Simplify.

$x = \dfrac{8 \pm 14}{2}$ $\sqrt{196} = 14$

$x = \dfrac{\boxed{}}{2}$ or $x = \dfrac{\boxed{}}{2}$ Write as two equations.

$= \boxed{}$ $= \boxed{}$ Simplify.

Your Turn Solve $x^2 + 13x = 0$ by using the Quadratic Formula.

EXAMPLE One Rational Root

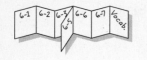

REMEMBER IT

The Quadratic Formula can be used to solve any quadratic equation.

2 Solve $x^2 - 34x + 289 = 0$ by using the Quadratic Formula.

Identify a, b, and c. Then, substitute these values into the Quadratic Formula.

$$x = \frac{-b \pm \sqrt{b^2 - 4ac}}{2a}$$ Quadratic Formula

$$x = \frac{-(-34) \pm \sqrt{}}{2(1)}$$ Replace a with 1, b with -34, and c with 289.

$$x = \frac{34 \pm \sqrt{0}}{2}$$ Simplify.

$$x = \frac{34}{2} \text{ or } \boxed{}$$ $\sqrt{0} = 0$

EXAMPLE Irrational Roots

FOLDABLES

ORGANIZE IT

On the page for Lesson 6-5, explain why factoring cannot be used to solve the quadratic equation in Example 3.

3 Solve $x^2 - 6x + 2 = 0$ by using the Quadratic Formula.

$$x = \frac{-b \pm \sqrt{b^2 - 4ac}}{2a}$$ Quadratic Formula

$$x = \frac{-(-6) \pm \sqrt{(-6)^2 - 4(1)(2)}}{2(1)}$$ Replace a with 1, b with -6, and c with 2.

$$x = \frac{6 \pm \sqrt{\boxed{}}}{2}$$ Simplify.

$$x = \frac{\boxed{}}{2} \text{ or } \frac{\boxed{}}{2}$$ $\sqrt{28} = \sqrt{4 \cdot 7}$ or $2\sqrt{7}$

The exact solutions are $\boxed{}$ and $\boxed{}$.

The approximate solutions are $\boxed{}$ and $\boxed{}$.

Your Turn Solve each equation by using the Quadratic Formula.

a. $x^2 - 22x + 121 = 0$

b. $x^2 - 5x + 3 = 0$

c. $x^2 + 5 = 4x$

EXAMPLE Describe Roots

KEY CONCEPT

Discriminant

Consider $ax^2 + bx + c = 0$.

- If $b^2 - 4ac > 0$ and $b^2 - 4ac$ is a perfect square, then there are 2 real, rational roots.

- If $b^2 - 4ac > 0$ and $b^2 - 4ac$ is *not* a perfect square, then there are 2 real, irrational roots.

- If $b^2 - 4ac = 0$, then there is one real, rational root.

- If $b^2 - 4ac < 0$, then there are two complex roots.

④ Find the value of the discriminant for each quadratic equation. Then describe the number and type of roots for the equation.

a. $x^2 + 6x + 9 = 0$

$a = \boxed{}, b = \boxed{}, c = \boxed{}$

$b^2 - 4ac = \boxed{}$

$= \boxed{}$

$= \boxed{}$

The discriminant is $\boxed{}$, so there $\boxed{}$.

b. $x^2 + 3x + 5 = 0$

$a = \boxed{}, b = \boxed{}, c = \boxed{}$

$b^2 - 4ac = \boxed{}$

$= \boxed{}$

$= \boxed{}$

The discriminant is $\boxed{}$, so there

$\boxed{}$.

Your Turn Find the value of the discriminant for each quadratic equation. Then describe the number and type of roots for the equation.

a. $x^2 + 8x + 16 = 0$

b. $x^2 + 2x + 7 = 0$

c. $x^2 + 3x + 1 = 0$

d. $x^2 + 4x - 12 = 0$

HOMEWORK ASSIGNMENT

Page(s):

Exercises:

Glencoe Algebra 2 **139**

Analyzing Graphs of Quadratic Functions

WHAT YOU'LL LEARN

- Analyze quadratic functions of the form $y = a(x - h)^2 + k$.

- Write a quadratic function in the form $y = a(x - h)^2 + k$.

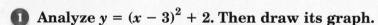

BUILD YOUR VOCABULARY (pages 124–125)

A function written in the form, $y = (x - h)^2 + k$, where

(h, k) is the [____] of the parabola and $x = h$ is its

[____] , is referred to as the **vertex form**.

EXAMPLE Graph a Quadratic Function in Vertex Form

1 Analyze $y = (x - 3)^2 + 2$. Then draw its graph.

The vertex is at (h, k) or [____] and the axis of symmetry is

$x =$ [____] . The graph has the same shape as the graph of

$y = x^2$, but is translated 3 units right and 2 units up.

Now use this information to draw the graph.

STEP 1 Plot the vertex, [____] .

STEP 2 Draw the axis of symmetry,
[____] .

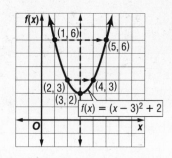

STEP 3 Find and plot two points
on one side of the axis of
symmetry, such as (2, 3)
and (1, 6).

STEP 4 Use symmetry to complete the graph.

FOLDABLES

ORGANIZE IT

On the page for
Lesson 6-6, sketch a
graph of a parabola.
Then sketch the graph
of the parabola after a
vertical translation and
a horizontal translation.

Your Turn Analyze $y = (x + 2)^2 - 4$. Then draw its graph.

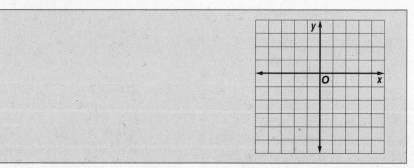

EXAMPLE Write $y = x^2 + bx + c$ in Vertex Form

2 Write $y = x^2 + 2x + 4$ in vertex form. Then analyze the function.

$y = x^2 + 2x + 4$ Notice that $x^2 + 2x + 4$ is not a perfect square.

$y =$ [   ]   Complete the square. Add $\left(\dfrac{2}{2}\right)^2$ or 1. Balance this addition by subtracting 1.

$y =$ [   ]   Write $x^2 + 2x + 1$ as a perfect square.

This function can be rewritten as $y = [x - (-1)]^2 + 3$.

So, $h =$ [   ] and $k =$ [   ].

The vertex is at [   ] and the axis of symmetry is

$x =$ [   ]. Since $a = 1$, the graph opens [   ] and has the

same shape as $y = x^2$ but is translated [   ] unit left and

[   ] units up.

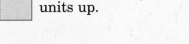

WRITE IT

Consider the vertex form of a quadratic function, $y = a(x - h)^2 + k$. Describe what happens to the graph as $|a|$ increases.

_____

_____

_____

_____

Your Turn Write each function in vertex form. Then analyze the function.

a. $y = x^2 + 6x + 5$

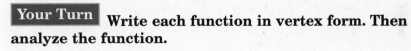

b. $y = -3x^2 - 6x + 4$

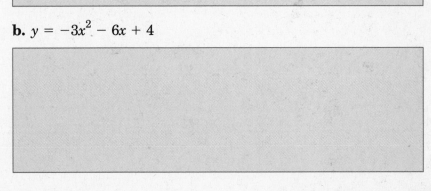

HOMEWORK ASSIGNMENT

Page(s): _____

Exercises: _____

_____

_____

Graphing and Solving Quadratic Inequalities

WHAT YOU'LL LEARN

- Graph quadratic inequalities in two variables.

- Solve quadratic inequalities in one variable.

FOLDABLES

ORGANIZE IT

On the page for Lesson 6-7, use your own words to describe the three steps for graphing quadratic inequalities.

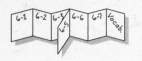

BUILD YOUR VOCABULARY (pages 124–125)

You can graph **quadratic inequalities** in two variables using the same techniques you used to graph _____ inequalities in two variables.

EXAMPLE Graph a Quadratic Inequality

1 Graph $y > x^2 - 3x + 2$.

STEP 1 Graph the related quadratic function, $y = x^2 - 3x + 2$. Since the inequality symbol is $>$, the parabola should be dashed.

$y = x^2 - 3x + 2$

STEP 2 Test a point inside the parabola, such as $(1, 2)$.

$$y > x^2 - 3x + 2$$

$2 > $ _____

$2 > $ _____

$2 > $ ____ ✔

So, $(1, 2)$ is a solution of the inequality.

STEP 3 Shade the region inside the parabola.

Your Turn Graph $y < -x^2 + 4x + 2$.

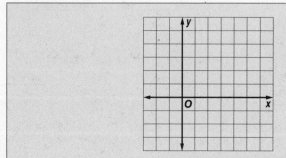

EXAMPLE Solve $ax^2 + bx + c > 0$

2 **Solve $x^2 - 4x + 3 > 0$ by graphing.**

The solution consists of the x values for which the graph of the related quadratic function lies *above* the x-axis. Begin by finding the roots of the related equation.

$$x^2 - 4x + 3 = 0 \qquad \text{Related equation}$$

$$\boxed{} = 0 \qquad \text{Factor.}$$

$$\boxed{} = 0 \text{ or } \boxed{} = 0 \qquad \text{Zero Product Property}$$

$$x = \boxed{} \qquad x = \boxed{} \qquad \text{Solve each equation.}$$

Sketch the graph of the parabola that has x-intercepts at 3 and 1. The graph lies above the x-axis to the left of $x = 1$ and to the right of $x = 3$.

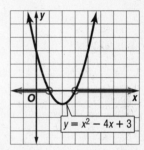

$y = x^2 - 4x + 3$

The solution set is $\boxed{}$.

Your Turn **Solve each inequality by graphing.**

a. $x^2 + 5x + 6 > 0$

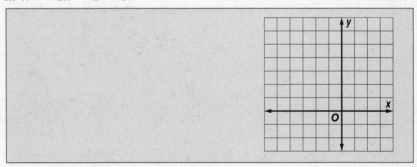

b. $x^2 + 6x + 2 \le 0$

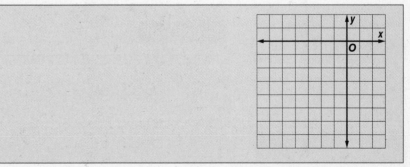

WRITE IT

Explain how you can check the solution set to a quadratic inequality.

_____

_____

_____

_____

_____

_____

EXAMPLE Solve a Quadratic Inequality

3 Solve $x^2 + x \leq 2$ algebraically.

First, solve the related equation $x^2 + x = 2$.

$$x^2 + x = 2 \qquad \text{Related quadratic equation}$$

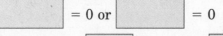 $= 0 \qquad$ Subtract 2 from each side.

$\boxed{} = 0 \qquad$ Factor.

$\boxed{} = 0$ or $\boxed{} = 0 \qquad$ Zero Product Property

$x = \boxed{} \qquad x = \boxed{} \qquad$ Solve each equation.

Plot the values on a number line. Use closed circles since these solutions are included. Notice that the number line is separated into 3 intervals.

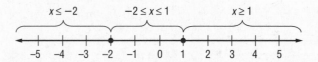

Test a value in each interval to see if it satisfies the original inequality.

$x \leq -2$	$-2 \leq x \leq 1$	$x \geq 1$
Test $x = -3$.	Test $x = 0$.	Test $x = 2$.
$x^2 + x \leq 2$	$x^2 + x \leq 2$	$x^2 + x \leq 2$
$(-3)^2 - 3 \overset{?}{\leq} 2$	$0^2 + 0 \overset{?}{\leq} 2$	$2^2 + 2 \overset{?}{\leq} 2$
$6 \leq 2$	$0 \leq 2$ ✔	$6 \leq 2$

The solution set is $\boxed{}$.

© Glencoe/McGraw-Hill

HOMEWORK ASSIGNMENT

Page(s):

Exercises:

Your Turn Solve $x^2 + 5x < -6$ algebraically.

BRINGING IT ALL TOGETHER

STUDY GUIDE

FOLDABLES™	VOCABULARY PUZZLEMAKER	BUILD YOUR VOCABULARY
Use your **Chapter 6 Foldable** to help you study for your chapter test.	To make a crossword puzzle, word search, or jumble puzzle of the vocabulary words in Chapter 6, go to: www.glencoe.com/sec/math/t_resources/free/index.php	You can use your completed **Vocabulary Builder** (pages 124–125) to help you solve the puzzle.

6-1 Graphing Quadratic Functions

Refer to the graph at the right as you complete the following sentences.

1. The curve is called a ⬚.

2. The line $x = -2$ is called the ⬚.

3. The point $(-2, 4)$ is called the ⬚.

Determine whether each function has a maximum or minimum value. Then find the maximum or minimum value of each function.

4. $f(x) = -x^2 + 2x + 5$

5. $f(x) = 3x^2 - 4x - 2$

6-2 Solving Quadratic Equations by Graphing

Solve each equation. If exact roots cannot be found, state the consecutive integers between which the roots are located.

6. $x^2 - 2x = 8$

7. $x^2 + 5x - 7 = 0$

6-3

Solving Quadratic Equations by Factoring

8. The solution of a quadratic equation by factoring is shown below. Give the reason for each step of the solution.

$x^2 - 10x = -21$ Original equation

$x^2 - 10x + 21 = 0$

$(x - 3)(x - 7) = 0$

$x - 3 = 0$ or $x - 7 = 0$

$x = 3$ $x = 7$

The solution set is [   ] .

Write a quadratic equation with the given roots. Write the equation in the form $ax^2 + bx + c = 0$, where a, b, and c are integers.

9. $-4, -2$

10. $3, 6$

6-4

Completing the Square

11. Solve $x^2 + 6x + 9 = 49$ by using the Square Root Property.

12. Solve $x^2 - 2x + 10 = 5$ by completing the square.

13. When the dimensions of a cube are reduced by 2 inches on each side, the surface area of the new cube is 486 square inches. What were the dimensions of the original cube?

6-5

The Quadratic Formula and the Discriminant

14. The value of the discriminant for a quadratic equation with integer coefficients is shown. Give the number and the type of roots for the equation.

Value of Discriminant	Number of Roots	Type of Roots
64		
−8		

15. Solve $x^2 - 8x = 2$ by using the Quadratic Formula. Find exact solutions.

6-6

Analyzing Graphs of Quadratic Functions

16. Match each graph with the description of the constants in the equation in vertex form.

a. $a > 0, h > 0, k < 0$ 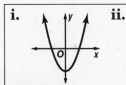   **b.** $a < 0, h < 0, k < 0$

c. $a < 0, h < 0, k > 0$ **d.** $a > 0, h = 0, k < 0$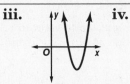

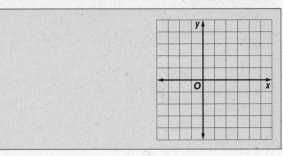

6-7

Graphing and Solving Quadratic Inequalities

17. Solve $0 < x^2 - 6x + 8$ by graphing.

Solve each inequality algebraically.

18. $x^2 - x > 20$

19. $x^2 - 10x < -16$

ARE YOU READY FOR THE CHAPTER TEST?

Visit **algebra2.com** to access your textbook, more examples, self-check quizzes, and practice tests to help you study the concepts in Chapter 6.

Check the one that applies. Suggestions to help you study are given with each item.

☐ **I completed the review of all or most lessons without using my notes or asking for help.**

• You are probably ready for the Chapter Test.

• You may want to take the Chapter 6 Practice Test on page 341 of your textbook as a final check.

☐ **I used my Foldable or Study Notebook to complete the review of all or most lessons.**

• You should complete the Chapter 6 Study Guide and Review on pages 336–340 of your textbook.

• If you are unsure of any concepts or skills, refer back to the specific lesson(s).

• You may also want to take the Chapter 6 Practice Test on page 341 of your textbook.

☐ **I asked for help from someone else to complete the review of all or most lessons.**

• You should review the examples and concepts in your Study Notebook and Chapter 6 Foldable.

• Then complete the Chapter 6 Study Guide and Review on pages 336–340 of your textbook.

• If you are unsure of any concepts or skills, refer back to the specific lesson(s).

• You may also want to take the Chapter 6 Practice Test on page 341 of your textbook.

Student Signature	Parent/Guardian Signature

Teacher Signature

Polynomial Functions

 Use the instructions below to make a Foldable to help you organize your notes as you study the chapter. You will see Foldable reminders in the margin of this Interactive Study Notebook to help you in taking notes.

Begin with five sheets of plain $8\frac{1}{2}$" × 11" paper.

STEP 1 **Stack and Fold**
Stack sheets of paper with edges $\frac{3}{4}$-inch apart. Fold up bottom edges to create equal tabs.

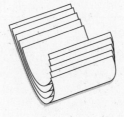

STEP 2 **Staple and Label**
Staple along fold. Label the tabs with lesson numbers.

 NOTE-TAKING TIP: When you take notes, preview the lesson and make generalizations about what you think you will learn. Then compare that with what you actually learned after each lesson.

7

BUILD YOUR VOCABULARY

This is an alphabetical list of new vocabulary terms you will learn in Chapter 7. As you complete the study notes for the chapter, you will see Build Your Vocabulary reminders to complete each term's definition or description on these pages. Remember to add the textbook page number in the second column for reference when you study.

Vocabulary Term	Found on Page	Definition	Description or Example
composition of functions			
depressed polynomial			
end behavior			
Factor Theorem			
Fundamental Theorem of Algebra			
inverse function			
inverse relation			
leading coefficients			
location principle			
one-to-one			

Vocabulary Term	Found on Page	Definition	Description or Example
polynomial function			
polynomial in one variable			
power function			
quadratic form			
Rational Zero Theorem			
relative maximum			
relative minimum			
remainder theorem			
square root function			
synthetic substitution [sihn-THEH-tihk]			

Polynomial Functions

WHAT YOU'LL LEARN

- Evaluate polynomial functions.

- Identify general shapes of graphs of polynomial functions.

The **leading coefficient** is the coefficient of the term with

the [ ] degree.

A common type of function is a **power function**, which has

an equation in the form [ ], where a and b are

real numbers.

KEY CONCEPT

A Polynomial in One Variable A polynomial of degree n in one variable x is an expression of the form

$a_0x^n + a_1x^{n-1} + \cdots + a_{n-2}x^2 + a_{n-1}x + a_n$,

where the coefficients $a_0, a_1, a_2, ..., a_n$, represent real numbers, a_0 is not zero, and n represents a nonnegative integer.

EXAMPLE Find Degree and Leading Coefficients

① State the degree and leading coefficient of each polynomial in one variable. If it is not a polynomial in one variable, explain why.

a. $7z^3 - 4z^2 + z$

This is a polynomial in one variable. The degree is [ ] and

the leading coefficient is [ ].

b. $6a^3 - 4a^2 + ab^2$

This [ ] a polynomial in one variable. It contains

two variables, [ ] and [ ].

Your Turn State the degree and leading coefficient of each polynomial in one variable. If it is not a polynomial in one variable, explain why.

a. $3x^3 + 2x^2 - 3$

b. $3x^2 + 2xy - 5$

EXAMPLE Evaluate a Polynomial Function

KEY CONCEPT

Definition of a Polynomial Function A polynomial function of degree n can be described by an equation of the form $P(x) = a_0x^n + a_1x^{n-1} + \cdots + a_{n-2}x^2 + a_{n-1}x + a_n$, where the coefficients $a_0, a_1, a_2, \ldots, a_n$, represent real numbers, a_0 is not zero, and n represents a nonnegative integer.

2 NATURE The total number of hexagons in a honeycomb can be modeled by the function $f(r) = 3r^2 - 3r + 1$, where r is the number of rings and $f(r)$ is the number of hexagons. Find the total number of hexagons in a honeycomb with 20 rings.

$f(r) = 3r^2 - 3r + 1$ Original function

$f(20) = $ Replace r with 20.

$= $ or   Simplify.

Your Turn Refer to Example 2. Find the total number of hexagons in a honeycomb with 30 rings.

EXAMPLE Functional Values of Variables

3 Find $p(y^3)$ if $p(x) = 2x^4 - x^3 + 3x$.

$p(x) = 2x^4 - x^3 + 3x$ Original function

$p(y^3) = $ Replace x with y^3.

$= $ Property of powers

Your Turn Find $f(x^3)$ if $f(a) = a^3 + 2a^2 + 2a$.

BUILD YOUR VOCABULARY (page 150)

The **end behavior** is the behavior of the graph as x approaches _____ $(+\infty)$ or _____ $(-\infty)$.

EXAMPLE Graphs of polynomial Functions

4 For each graph, describe the end behavior, determine whether it represents an odd-degree or an even-degree function, and state the number of real zeros.

a.

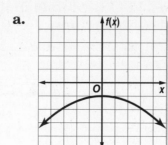

- $f(x) \rightarrow$ [ ] as $x \rightarrow +\infty$.

- $f(x) \rightarrow$ [ ] as $x \rightarrow -\infty$.

- It is an [ ] polynomial function.

- The function has [ ] real zeros.

b.

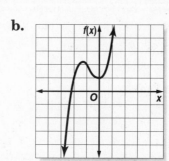

- $f(x) \rightarrow$ [ ] as $x \rightarrow +\infty$.

- $f(x) \rightarrow$ [ ] as $x \rightarrow -\infty$.

- It is an [ ] polynomial function.

- The function has [ ] real zero.

Your Turn For each graph, describe the end behavior, determine whether it represents an odd-degree or an even-degree function, and state the number of real zeros.

a.

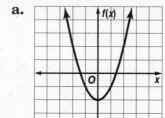

b.

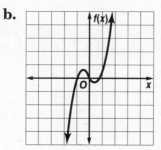

Graphing Polynomial Functions

WHAT YOU'LL LEARN

- Graph polynomial functions and locate their real zeros.

- Find the maxima and minima of polynomial functions.

FOLDABLES™

ORGANIZE IT

On the tab for Lesson 7-2, explain how knowing the end behavior of a graph will assist you in completing the sketch of the graph.

Polynomials
7-1
7-2
7-3
7-4
7-5
7-6
7-7
7-8
7-9

EXAMPLE Graph a Polynomial Function

❶ **Graph $f(x) = -x^3 - 4x^2 + 5$ by making a table of values.**

This is an odd degree polynomial with a negative leading coefficient, so

$f(x) \rightarrow$ ☐ as $x \rightarrow -\infty$ and

$f(x) \rightarrow$ ☐ as $x \rightarrow +\infty$.

The graph intersects the x-axis at ☐ points indicating that there are ☐ real zeros.

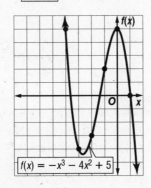

$f(x) = -x^3 - 4x^2 + 5$

x	$f(x)$
-4	
-3	
-2	
-1	
0	
1	
2	

Your Turn Graph $f(x) = x^3 + 2x^2 + 1$ by making a table of values.

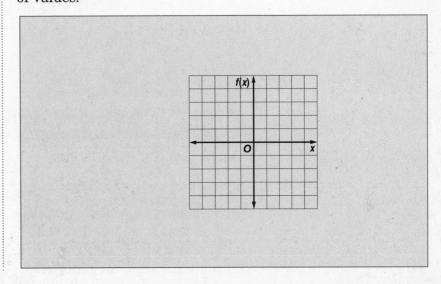

EXAMPLE Locate Zeros of a Function

2 Determine consecutive values of x between which each real zero of the function $f(x) = x^4 - x^3 - 4x^2 + 1$ is located. Then draw the graph.

Make a table of values. Since $f(x)$ is a 4th degree polynomial function, it will have between 0 and 4 zeros, inclusive. Look at the value of $f(x)$ to locate the zeros. Then use the points to sketch the graph of the function.

KEY CONCEPT

Location Principle
Suppose $y = f(x)$ represents a polynomial function and a and b are two numbers such that $f(a) < 0$ and $f(b) > 0$. Then the function has at least one real zero between a and b.

x	$f(x)$
-2	9
-1	-1
0	1
1	-3
2	-7
3	19

} sign change

} sign change

} sign change

} sign change

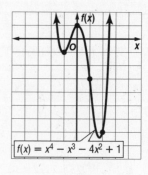

$f(x) = x^4 - x^3 - 4x^2 + 1$

There are zeros between $x =$ ☐ and ☐, $x =$ ☐ and ☐, $x =$ ☐ and ☐, and $x =$ ☐ and ☐.

Your Turn Determine consecutive values of x between which each real zero of the function $f(x) = x^3 - 4x^2 + 2$ is located. Then draw the graph.

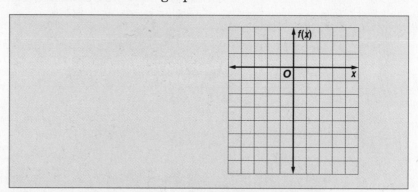

BUILD YOUR VOCABULARY (page 151)

A point on a graph is a **relative maximum** of a function if no other nearby points have a ☐ y-coordinate. Likewise, a point is a **relative minimum** if no other nearby points have a ☐ y-coordinate.

EXAMPLE **Maximum and Minimum Points**

3 Graph $f(x) = x^3 - 4x^2 + 5$. Estimate the x-coordinates at which the relative maximum and relative minimum occur.

Make a table of values and graph the function.

$f(x) = x^3 - 4x^2 + 5$

x	f(x)
−2	−19
−1	0
0	5
1	2
2	−3
3	−4
4	5
5	30

zero at $x = -1$

◄ indicates a relative maximum

} zero between $x = 1$ and $x = 2$

◄ indicates a relative minimum

} and zero between $x = 3$ and $x = 4$

The value of $f(x)$ at $x =$ ☐ is greater than the surrounding

points, so it is a relative ☐ . The value of $f(x)$ at

about $x =$ ☐ is less than the surrounding points, so it is a

relative ☐ .

Your Turn Graph $f(x) = x^3 + 3x^2 + 2$. Estimate the x-coordinates at which the relative maximum and relative minimum occur.

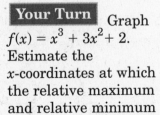

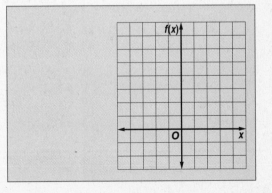

HOMEWORK ASSIGNMENT

Page(s):

Exercises:

Solving Equations Using Quadratic Techniques

EXAMPLE Write an Expression in Quadratic Form

WHAT YOU'LL LEARN

- Write expressions in quadratic form.
- Use quadratic techniques to solve equations.

1 Write each expression in quadratic form, if possible.

a. $2x^6 + x^3 + 9$ $\boxed{} + (x^3) + 9$ $x^6 = \boxed{}$

b. $x^4 + 2x^3 + 1$ This cannot be written in quadratic form since

$$x^4 \neq \boxed{}.$$

c. $x^{\frac{2}{3}} + 2x^{\frac{1}{3}} - 4$ $\boxed{} + 2x^{\frac{1}{3}} - 4$ $x^{\frac{2}{3}} = \boxed{}$

KEY CONCEPT

Quadratic Form An expression that is quadratic in form can be written as $au^2 + bu + c$ for any numbers a, b, and c, $a \neq 0$, where u is some expression in x. The expression $au^2 + bu + c$ is called the quadratic form of the original expression.

Your Turn Write each expression in quadratic form, if possible.

a. $2x^4 + x^2 + 3$

b. $x^6 + x^4 + 1$

c. $x - 2x^{\frac{1}{2}} + 3$

d. $x^{12} + 5$

EXAMPLE Solve Polynomial Equations

2 a. Solve $x^4 - 29x^2 + 100 = 0$.

$$x^4 - 29x^2 + 100 = 0 \quad \text{Original equation}$$

$\boxed{} - 29(x^2) + 100 = 0$ Write the expression on the left in quadratic form.

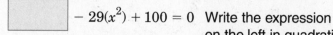 $= 0$ Factor the trinomial.

 $= 0$ Factor.

© Glencoe/McGraw-Hill

Use the Zero Product Property.

$$\boxed{} = 0 \qquad \text{or} \qquad \boxed{} = 0$$

$$x = \boxed{} \qquad\qquad\qquad x = \boxed{}$$

$$\boxed{} = 0 \qquad \text{or} \qquad \boxed{} = 0$$

$$x = \boxed{} \qquad\qquad\qquad x = \boxed{}$$

b. Solve $x^3 + 216 = 0$.

$$x^3 + 216 = 0 \qquad \text{Original equation}$$

$$x^3 + \boxed{} = 0 \qquad \text{Sum of two cubes}$$

$$(x + 6)\boxed{} = 0 \qquad \text{Factor.}$$

$$\boxed{} = 0 \text{ or } \boxed{} = 0 \qquad \text{Zero Product Property}$$

The solution of the first equation is $\boxed{}$. The second equation can be solved by using the Quadratic Formula.

$$x = \frac{-b \pm \sqrt{b^2 - 4ac}}{2a} \qquad \text{Quadratic Formula}$$

$$x = \frac{-(-6) \pm \sqrt{(-6)^2 - 4(1)(36)}}{2(1)} \qquad \begin{array}{l}\text{Replace } a \text{ with 1,} \\ b \text{ with } -6, \text{ and} \\ c \text{ with 36.}\end{array}$$

$$x = \frac{6 \pm \boxed{}}{2} \qquad \text{Simplify.}$$

$$x = \frac{6 \pm \boxed{}}{2} \quad \text{or} \quad \frac{6 \pm 6i\sqrt{3}}{2} \qquad \begin{array}{l}\text{Multiply } \sqrt{108} \text{ and} \\ \sqrt{-1}\end{array}$$

$$x = \boxed{} \qquad \text{Simplify.}$$

Your Turn Solve each equation.

a. $x^4 - 10x^2 + 9 = 0$ **b.** $x^3 + 8 = 0$

FOLDABLES

ORGANIZE IT

Take notes, define terms, record concepts, and write examples from this lesson under the 7-3 tab.

EXAMPLE Solve Equations with Rational Exponents

3 Solve $x^{\frac{1}{2}} - x^{\frac{1}{4}} - 6 = 0$.

$$x^{\frac{1}{2}} - x^{\frac{1}{4}} - 6 = 0 \qquad \text{Original equation}$$

$$\boxed{} - \left(x^{\frac{1}{4}}\right) - 6 = 0 \qquad \text{Write the expression on the left in quadratic form.}$$

$$\boxed{} = 0 \qquad \text{Factor the trinomial.}$$

$$\boxed{} = 0 \text{ or } \boxed{} = 0 \qquad \text{Zero Product Property}$$

$$\boxed{} = 3 \qquad \boxed{} = -2 \qquad \text{Isolate } x.$$

$$\boxed{} = 3^4 \qquad \boxed{} = (-2)^4 \qquad \text{Raise each side to the fourth power.}$$

$$x = \boxed{} \qquad x = \boxed{}$$

By substituting each value into the original equation, you find that $\boxed{}$ is the only solution.

Your Turn Solve $x^{\frac{2}{3}} + 5x^{\frac{1}{3}} + 6 = 0$.

HOMEWORK ASSIGNMENT

Page(s):

Exercises:

The Remainder and Factor Theorems

WHAT YOU'LL LEARN

- Evaluate functions using synthetic substitution.
- Determine whether a binomial is a factor of a polynomial by using synthetic substitution.

KEY CONCEPTS

Remainder Theorem
If a polynomial $f(x)$ is divided by $x - a$, the remainder is the constant $f(a)$, and

$$\underbrace{f(x)}_{\text{Dividend}} \underbrace{=}_{\text{equals}} \underbrace{q(x)}_{\text{quotient}} \cdot \underbrace{(x - a)}_{\substack{\text{times divisor}}} + \underbrace{f(a).}_{\substack{\text{plus remainder.}}}$$

where $q(x)$ is a polynomial with degree one less than the degree of $f(x)$.

Factor Theorem
The binomial $x - a$ is a factor of the polynomial $f(x)$ if and only if $f(a) = 0$.

BUILD YOUR VOCABULARY (page 151)

When synthetic division is used to evaluate a function, it is called **synthetic substitution**.

EXAMPLE Synthetic Substitution

1 If $f(x) = 3x^4 - 2x^3 + x^2 - 2$, find $f(4)$.

METHOD 1 Synthetic Substitution

By the Remainder Theorem, $f(4)$ should be the remainder when you divide the polynomial by $x - 4$.

4	3	−2	1	0	−2

Notice that there is no x term. A zero is placed in this position as a placeholder.

The remainder is _____ . Thus, by using synthetic substitution, $f(4) =$ _____ .

METHOD 2 Direct Substitution

Replace x with 4.

$f(x) = 3x^4 - 2x^3 + x^2 - 2$ Original function

$f(4) =$ _____   Replace x with 4.

$f(4) =$ _____ or _____   Simplify.

By using direct substitution, $f(4) =$ _____ .

Your Turn If $f(x) = 2x^3 - 3x^2 + 7$, find $f(3)$.

When you divide a polynomial by one of its **binomial** factors, the quotient is called a **depressed polynomial**.

EXAMPLE Use the Factor Theorem

2 Show that $x - 3$ is a factor of $x^3 + 4x^2 - 15x - 18$. Then find the remaining factors of the polynomial.

The binomial $x - 3$ is a factor of the polynomial if 3 is a zero of the related polynomial function. Use the factor theorem and synthetic division.

$$\underline{3}| \qquad 1 \qquad 4 \qquad -15 \qquad -18$$

Since the remainder is 0, [ ] is a factor of the polynomial. The polynomial $x^3 + 4x^2 - 15x - 18$ can be factored as [ ]. The polynomial

[ ] is the depressed polynomial. Check to see if this polynomial can be factored.

$x^2 + 7x + 6 = $ [ ]   Factor the trinomial.

So, $x^3 + 4x^2 - 15x - 18 = $ [ ].

Your Turn Show that $x + 2$ is a factor of $x^3 + 8x^2 + 17x + 10$. Then find the remaining factors of the polynomial.

FOLDABLES

ORGANIZE IT

On the tab for Lesson 7-4, use the Factor Theorem to show that $x + 3$ is a factor of $3x^3 - 3x^2 - 36x$.

Polynomials
7-1
7-2
7-3
7-4
7-5
7-6
7-7
7-8
7-9

HOMEWORK ASSIGNMENT

Page(s):

Exercises:

Roots and Zeros

WHAT YOU'LL LEARN

- Determine the number and type of roots for a polynomial equation.

- Find the zeros of a polynomial function.

KEY CONCEPTS

Fundamental Theorem of Algebra Every polynomial equation with degree greater than zero has at least one root in the set of complex numbers.

Corollary A polynomial equation of the form $P(x) = 0$ of degree n with complex coefficients has exactly n roots in the set of complex numbers.

FOLDABLES On the tab for Lesson 7-5, write these key concepts. Be sure to include examples.

EXAMPLE Determine Number and Type of Roots

1 Solve each equation. State the number and type of roots.

a. $a - 10 = 0$

$a - 10 = 0$ Original equation

$a = \boxed{}$ Add 10 to each side.

This equation has exactly one real root, $\boxed{}$.

b. $x^2 + 2x - 48 = 0$

$x^2 + 2x - 48 = 0$ Original equation

$\boxed{} = 0$ Factor.

Use the Zero Product Property.

$\boxed{} = 0$ or $\boxed{} = 0$

$x = \boxed{}$ $x = \boxed{}$ Solve each equation.

This equation has two real roots, $\boxed{}$ and $\boxed{}$.

Your Turn Solve each equation. State the number and type of roots.

a. $x + 3 = 0$

b. $x^2 - x - 12 = 0$

c. $a^4 - 81 = 0$

EXAMPLE **Find Numbers of Positive and Negative Zeros**

KEY CONCEPT

Descartes' Rule of Signs
If $P(x)$ is a polynomial with real coefficients whose terms are arranged in descending powers of the variable,

- the number of positive real zeros of $y = P(x)$ is the same as the number of changes in sign of the coefficients of the terms, or is less than this by an even number, and

- the number of negative real zeros of $y = P(x)$ is the same as the number of changes in sign of the coefficients of the terms of $P(-x)$, or is less than this number by an even number.

② State the possible number of positive real zeros, negative real zeros, and imaginary zeros of $p(x) = -x^6 + 4x^3 - 2x^2 - x - 1$.

Since $p(x)$ has degree 6, it has 6 zeros. However, some of them may be imaginary. Use Descartes' Rule of Signs to determine the number and type of real zeros. Count the number of changes in sign for the coefficients of $p(x)$.

$$p(x) = \quad -x^6 \quad + \quad 4x^3 \quad - \quad 2x^2 \quad - \quad x \quad - \quad 1$$

Since there are two sign changes, there are 2 or 0 positive real zeros. Find $p(-x)$ and count the number of sign changes for its coefficients.

$$p(-x) = -(-x)^6 + 4(-x)^3 - 2(-x)^2 - (-x) - 1$$
$$= -x^6 \quad - \quad 4x^3 \quad - \quad 2x^2 \quad + \quad x \quad - \quad 1$$

Since there are two sign changes, there are 2 or 0 negative real zeros. Make a chart of possible combinations.

Positive Real Zeros	Negative Real Zeros	Imaginary Zeros	Total
2	2	2	6
0	2	4	6
2	0	4	6
0	0	6	6

HOMEWORK ASSIGNMENT

Page(s):

Exercises:

Your Turn State the possible number of positive real zeros, negative real zeros, and imaginary zeros of $p(x) = x^4 - x^3 + x^2 + x + 3$.

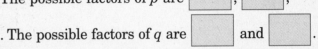

Rational Zero Theorem

WHAT YOU'LL LEARN

- Identify the possible rational zeros of a polynomial function.
- Find all the rational zeros of a polynomial function.

KEY CONCEPTS

Rational Zero Theorem
Let $f(x) =$

$a_0x^n + a_1x^{n-1} + \cdots +$

$a_{n-2}x^2 + a_{n-1}x + a_n$

represent a polynomial function with integral coefficients. If $\frac{p}{q}$ is a rational number in simplest form and is a zero of $y = f(x)$, then p is a factor of a_n and q is a factor of a_0.

Corollary (Integral Zero Theorem) If the coefficients of a polynomial function are integers such that $a_0 = 1$ and $a_n \neq 0$, any rational zeros of the function must be factors of a_n.

EXAMPLE Identify Possile Zeros

1 List all of the possible rational zeros of each function.

a. $f(x) = 3x^4 - x^3 + 4$

If $\frac{p}{q}$ is a rational zero, then p is a factor of 4 and q is a factor of 3. The possible factors of p are ⬚ , ⬚ , and ⬚ . The possible factors of q are ⬚ and ⬚ .

So, $\frac{p}{q} =$ ⬚ .

b. $f(x) = x^4 + 7x^3 - 15$

Since the coefficient of x^4 is ⬚ the possible rational zeros must be the factors of the constant term -15. So, the possible rational zeros are ⬚ .

Your Turn List all of the possible rational zeros of each function.

a. $f(x) = 2x^3 + x + 6$

⬚

b. $f(x) = x^3 + 3x + 24$

⬚

EXAMPLE Find All Zeros

2 Find all of the zeros of $f(x) = x^4 + x^3 - 19x^2 + 11x + 30$.

From the corollary to the Fundamental Theorem of Algebra, we know there are exactly 4 complex roots.

According to Descartes' Rule of Signs, there are 2 or 0 positive real roots and 2 or 0 negative real roots.

The possible rational zeros are ± 1, ± 2, ± 3, ± 5, ± 6, ± 10, ± 15, and ± 30.

ORGANIZE IT

On the tab for Lesson 7-6, write how you would find the possible rational zeros of $f(x) = x^3 + 6x^2 + x + 6$.

Make a table and test some possible rational zeros.

$\frac{p}{q}$	1	1	−19	11	30
0					
1					
2					

Since $f(2) = 0$, you know that $x = 2$ is a zero. The depressed polynomial is _____ .

Since $x = 2$ is a positive real zero, and there can only be 2 or 0 positive real zeros, there must be one more positive real zero. Test the next possible rational zeros on the depressed polynomial.

$\frac{p}{q}$	1	3	−13	−15
3				

There is another zero at $x = 3$. The depressed polynomial is _____ .

Factor $x^2 + 6x + 5$.

$$x^2 + 6x + 5 = 0$$ Write the depressed polynomial.

_____ = 0 Factor.

_____ = 0 or _____ = 0 Zero Product Property

$x = $ _____ $x = $ _____

There are two more real zeros at $x = $ _____ and $x = $ _____ .

Your Turn Find all of the zeros of $f(x) = x^4 + 4x^3 - 14x^2 - 36x + 45$.

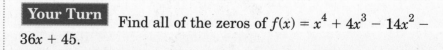

HOMEWORK ASSIGNMENT

Page(s): _____

Exercises: _____

Operations on Functions

Add and Subtract Functions

WHAT YOU'LL LEARN

- Find the sum, difference, product, and quotient of functions.

- Find the composition of functions.

① Given $f(x) = 3x^2 + 7x$ and $g(x) = 2x^2 - x - 1$, find $(f + g)(x)$.

$(f + g)(x) = f(x) + g(x)$

$$= \boxed{} + \boxed{}$$

$$= \boxed{}$$

KEY CONCEPT

Operations with Functions

Sum
$(f + g)(x) = f(x) + g(x)$

Difference
$(f - g)(x) = f(x) - g(x)$

Product
$(f \cdot g)(x) = f(x) \cdot g(x)$

Quotient
$\left(\dfrac{f}{g}\right)(x) = \dfrac{f(x)}{g(x)}, g(x) \neq 0$

Your Turn Given $f(x) = 2x^2 + 5x + 2$ and $g(x) = 3x^2 + 3x - 4$, find each function.

a. $(f + g)(x)$

b. $(f - g)(x)$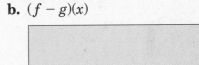

Multiply and Divide Functions

② Given $f(x) = 3x^2 - 2x + 1$ and $g(x) = x - 4$, find $(f \cdot g)(x)$.

$(f \cdot g)(x) = f(x) \cdot g(x)$

$$= \left(\boxed{}\right)\left(\boxed{}\right)$$

$= 3x^2(x - 4) - 2x(x - 4) + 1(x - 4)$ Distributive Property

$$= \boxed{}$$ Distributive Property

$$= \boxed{}$$ Simplify.

Your Turn Given $f(x) = 2x^2 + 3x - 1$ and $g(x) = x + 2$, find each function.

a. $(f \cdot g)(x)$

b. $\left(\dfrac{f}{g}\right)(x)$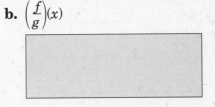

EXAMPLE Simplify Composition of Functions

Composition of Functions Suppose f and g are functions such that the range of g is a subset of the domain of f. Then the composite function $f \circ g$ can be described by the equation $[f \circ g](x) = f[g(x)]$.

FOLDABLES On the tab for Lesson 7-7, write how you would read $[f \circ g](x)$. Then explain which function, f or g, you would evaluate first.

3 **a.** Find $[f \circ g](x)$ and $[g \circ f](x)$ for $f(x) = 3x^2 - x + 4$ and $g(x) = 2x - 1$.

$[f \circ g](x) = f[g(x)]$

$= f\left(\boxed{}\right)$

$= \boxed{} - \boxed{} + \boxed{}$

$= 3(4x^2 - 4x + 1) - 2x + 1 + 4$

$= \boxed{}$

$[g \circ f](x) = g(f(x))$

$= g\left(\boxed{}\right)$

$= \boxed{}\,(3x^2 - x + 4) - \boxed{}$

$= \boxed{}$

b. Evaluate $[f \circ g](x)$ and $[g \circ f](x)$ for $x = -2$.

$[f \circ g](x) = 12x^2 - 14x + 8$

$[f \circ g](-2) = \boxed{}$

$= \boxed{}$

$[g \circ f](x) = 6x^2 - 2x + 7$

$[g \circ f](-2) = \boxed{}$

$= \boxed{}$

Your Turn

a. Find $[f \circ g](x)$ and $[g \circ f](x)$ for $f(x) = x^2 + 2x + 3$ and $g(x) = x + 5$.

$\boxed{}$

b. Evaluate $[f \circ g](x)$ and $[g \circ f](x)$ for $x = 1$.

$\boxed{}$

Inverse Functions and Relations

WHAT YOU'LL LEARN

- Find the inverse of a function or relation.

- Determine whether two functions or relations are inverses.

KEY CONCEPTS

Inverse Relations Two relations are inverse relations if and only if whenever one relation contains the element (a, b), the other relation contains the element (b, a).

Property of Inverse Functions Suppose f and f^{-1} are inverse functions. Then, $f(a) = b$ if and only if $f^{-1}(b) = a$.

EXAMPLE Find an Inverse Relation

1 **GEOMETRY** The ordered pairs of the relation {(1, 3), (6, 3), (6, 0), (1, 0)} are the coordinates of the vertices of a rectangle. Find the inverse of this relation and determine whether the resulting ordered pairs are also the coordinates of the vertices of a rectangle.

To find the inverse of this relation, reverse the coordinates of the ordered pairs. The inverse of the relation is

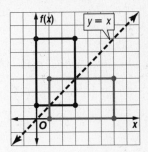

Plotting the points shows that the ordered pairs also describe the vertices of a rectangle. Notice that the graph of the relation and the inverse are reflections over the graph of $y = x$.

EXAMPLE Find an Inverse Function

2 Find the inverse of $f(x) = -\dfrac{1}{2}x + 1$.

STEP 1 Replace $f(x)$ with y in the original equation.

$$f(x) = -\frac{1}{2}x + 1 \quad \longrightarrow \quad \boxed{}$$

STEP 2 Interchange x and y.

$$\boxed{}$$

STEP 3 Solve for y.

$$x = -\frac{1}{2}y + 1 \qquad \text{Inverse}$$

$$\boxed{} = \boxed{} \qquad \text{Multiply each side by } -2.$$

$$\boxed{} = \boxed{} \qquad \text{Add 2 to each side.}$$

STEP 4 Replace y with $f^{-1}(x)$.

$$y = \boxed{} \quad \longrightarrow \quad f^{-1}(x) = \boxed{}$$

EXAMPLE Verify Two Functions are Inverses

3 Determine whether $f(x) = \frac{3}{4}x - 6$ and $g(x) = \frac{4}{3}x + 8$ are inverse functions.

Check to see if the compositions of $f(x)$ and $g(x)$ are identity functions.

$[f \circ g](x)$ $\qquad\qquad\qquad$ $[g \circ f](x)$

$= f\left(\boxed{}\right)$ $\qquad\qquad$ $= g\left(\boxed{}\right)$

$= \boxed{}\left(\frac{4}{3}x + 8\right) - \boxed{}$ $\qquad$ $= \boxed{}\left(\frac{3}{4}x - 6\right) + \boxed{}$

$= \boxed{}$ $\qquad\qquad\qquad$ $= \boxed{}$

$= \boxed{}$ $\qquad\qquad\qquad\qquad$ $= \boxed{}$

The functions are inverses since both compositions equal $\boxed{}$.

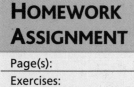

Your Turn

a. The ordered pairs of the relation $\{(-3, 4), (-1, 5), (2, 3), (1, 1), (-2, 1)\}$ are the coordinates of the vertices of a pentagon. Find the inverse of this relation and determine whether the resulting ordered pairs are also the coordinates of the vertices of a pentagon.

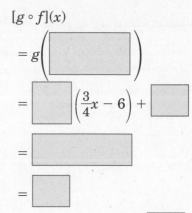

b. Find the inverse of $f(x) = \frac{1}{3}x + 6$.

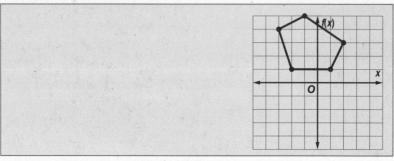

c. Determine whether $f(x) = 3x + 1$ and $g(x) = \frac{x-1}{3}$ are inverse functions.

HOMEWORK ASSIGNMENT

Page(s):

Exercises:

Square Root Functions and Inequalities

WHAT YOU'LL LEARN

- Graph and analyze square root functions.

- Graph square root inequalities.

EXAMPLE Graph a Square Root Function

1 **Graph $y = \sqrt{\dfrac{3}{2}x - 1}$. State the domain, range, and x- and y-intercepts.**

Since the radicand cannot be negative, identify the domain.

$$\boxed{} \geq 0 \qquad \text{Write the expression inside the radicand as } \geq 0.$$

$$x \geq \boxed{} \qquad \text{Solve for } x.$$

The x-intercept is $\boxed{}$.

Make a table of values to graph the function. The domain is $x \geq \dfrac{2}{3}$ and the range is $y \geq 0$.

The x-intercept is $\dfrac{2}{3}$. There is no y-intercept.

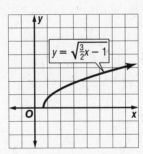

x	y
$\dfrac{2}{3}$	0
1	0.71
2	1.41
3	1.87
4	2.24
5	2.55
6	2.83

Your Turn Graph $y = \sqrt{2x - 2}$. State the domain, range, and x- and y-intercepts.

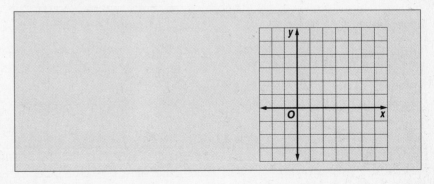

Graph a Square Root Inequality

2 **a.** **Graph** $y > \sqrt{3x + 5}$.

Graph the related equation $y = \sqrt{3x + 5}$. Since the boundary is not included, the graph should be dashed.

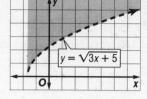

The domain includes values for $x \geq -\dfrac{5}{3}$. So the graph is to the

right of $x = -\dfrac{5}{3}$.

Select a point and test its ordered pair. Test $(0, 0)$.

$0 > \boxed{}$

$0 > \boxed{}$ false

Shade the region that does not include $(0, 0)$.

b. **Graph** $y \leq \sqrt{4 + \dfrac{3}{2}x}$.

Graph the related equation $y = \sqrt{4 + \dfrac{3}{2}x}$.

The domain includes values for

$x \geq -\dfrac{8}{3}$. Test $(4, 1)$.

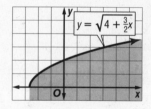

$1 \leq \boxed{}$

$1 \leq \boxed{}$ true

Shade the region that includes $(4, 1)$.

Your Turn   Graph $y \leq \sqrt{3 + 3x}$.

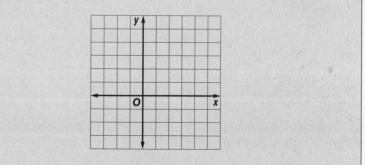

HOMEWORK ASSIGNMENT

Page(s):

Exercises:

STUDY GUIDE

FOLDABLES™	**VOCABULARY PUZZLEMAKER**	**BUILD YOUR VOCABULARY**
Use your **Chapter 7 Foldable** to help you study for your chapter test.	To make a crossword puzzle, word search, or jumble puzzle of the vocabulary words in Chapter 7, go to: www.glencoe.com/sec/math/ t_resources/free/index.php	You can use your completed **Vocabulary Builder** (pages 150–151) to help you solve the puzzle.

7-1

Polynomial Functions

1. Give the degree and leading coefficient of each polynomial.

degree   **leading coefficient**

a. $10x^3 + 3x^2 - x + 7$

b. $7y^2 - 2y^5 + y - 4y^3$

c. 100

7-2

Graphing Polynomial Functions

2. Graph $f(x) = x^3 - 6x^2 + 2x + 8$ by making a table of values. Let $x = \{-2, -1, 0, 1, 2, 3\}$. Then determine consecutive values of x between which each real zero is located. Estimate the x-coordinates at which the relative maxima and relative minima occur.

© Glencoe/McGraw-Hill

7-3

Solving Equations Using Quadratic Techniques

3. Which of the following expressions can be written in quadratic form?

   a. $x^3 + 6x^2 + 9$ b. $x^4 - 7x^2 + 6$ c. $m^6 + 4m^3 + 4$

   d. $y - 2y^{\frac{1}{2}} - 15$ e. $x^5 + x^3 + 1$ f. $r^4 + 6 - r^3$

4. Solve $x^3 + 2x^2 = 15x$.

7-4

The Remainder and Factor Theorems

Find $f(-2)$ for each function.

5. $f(x) = x^3 + 4x^2 - 8x - 6$

6. $f(x) = x^3 + 4x^2 + 4x$

7-5

Roots and Zeros

Let $f(x) = x^6 - 2x^5 + 3x^4 - 4x^3 + 5x^2 + 6x - 7$.

7. What are the possible numbers of positive real zeros of f?

8. Write $f(-x)$ in simplified form (with no parentheses).

9. What are the possible numbers of negative real zeros of f?

7-6

Rational Zero Theorem

10. List all the possible values of p, all the possible values of q, and all the possible rational zeros $\dfrac{p}{q}$ for $f(x) = x^3 - 2x^2 - 11x + 12$.

    possible values of p:

    possible values of q:

    possible values of $\dfrac{p}{q}$:

7-7

Operations on Functions

11. Find $(f + g)(x)$, $(f - g)(x)$, $(f \cdot g)(x)$, and $\left(\dfrac{f}{g}\right)(x)$ for

$f(x) = x^2 - 5x + 2$ and $g(x) = 3x + 6$.

12. Find $[g \circ h](x)$ and $[h \circ g](x)$ for $g(x) = x^2 + 8x - 7$ and $h(x) = x - 3$.

7-8

Inverse Functions and Relations

13. Find the inverse of the function $f(x) = -5x + 4$.

14. Determine whether $g(x) = 2x + 4$ and $f(x) = \dfrac{x}{2} + 2$ are inverse functions.

7-9

Square Root Functions and Inequalities

15. Graph $y < 2 + \sqrt{3x + 4}$. Then state the domain and range of the function.

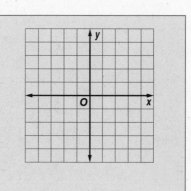

ARE YOU READY FOR THE CHAPTER TEST?

Math Online

Visit **algebra2.com** to access your textbook, more examples, self-check quizzes, and practice tests to help you study the concepts in Chapter 7.

Check the one that applies. Suggestions to help you study are given with each item.

☐ **I completed the review of all or most lessons without using my notes or asking for help.**

- You are probably ready for the Chapter Test.

- You may want to take the Chapter 7 Practice Test on page 405 of your textbook as a final check.

☐ **I used my Foldable or Study Notebook to complete the review of all or most lessons.**

- You should complete the Chapter 7 Study Guide and Review on pages 400–404 of your textbook.

- If you are unsure of any concepts or skills, refer back to the specific lesson(s).

- You may also want to take the Chapter 7 Practice Test on page 405.

☐ **I asked for help from someone else to complete the review of all or most lessons.**

- You should review the examples and concepts in your Study Notebook and Chapter 7 Foldable.

- Then complete the Chapter 7 Study Guide and Review on pages 400–404 of your textbook.

- If you are unsure of any concepts or skills, refer back to the specific lesson(s).

- You may also want to take the Chapter 7 Practice Test on page 405.

Student Signature	Parent/Guardian Signature

Teacher Signature

Conic Sections

 Use the instructions below to make a Foldable to help you organize your notes as you study the chapter. You will see Foldable reminders in the margin of this Interactive Study Notebook to help you in taking notes.

Begin with four sheets of grid paper and one piece of construction paper.

STEP 1 **Fold and Staple**
Stack sheets of grid papr with edges $\frac{1}{2}$ inch apart. Fold top edges back. Staple to construction paper at top.

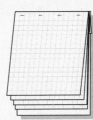

STEP 2 **Cut and Label**
Cut grid paper in half lengthwise. Label tabs as shown.

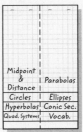

NOTE-TAKING TIP: Remember to always take notes on your own. Don't use someone else's notes as they may not make sense.

BUILD YOUR VOCABULARY

This is an alphabetical list of new vocabulary terms you will learn in Chapter 8. As you complete the study notes for the chapter, you will see Build Your Vocabulary reminders to complete each term's definition or description on these pages. Remember to add the textbook page number in the second column for reference when you study.

Vocabulary Term	Found on Page	Definition	Description or Example
asymptote [A-suhm(p)-TOHT]			
center of a circle			
center of an ellipse			
circle			
conic section			
conjugate axis [KAHN-jih-guht]			
directrix [duh-REHK-trihks]			
distance formula			
ellipse [ih-LIHPS]			

Vocabulary Term	Found on Page	Definition	Description or Example
foci of an ellipse			
focus of a parabola [FOH-kuhs]			
hyperbola [hy-PUHR-buh-luh]			
latus rectum [LA-tuhs REHK-tuhm]			
major axis			
midpoint formula			
minor axis			
parabola [puh-RA-buh-luh]			
tangent [TAN-juhnt]			
transverse axis			

Midpoint and Distance Formulas

EXAMPLE Find a Midpoint

1 COMPUTERS A graphing program draws a line segment on a computer screen so that its ends are at (5, 2) and (7, 8). What are the coordinates of its midpoint?

$$\left(\frac{x_1 + x_2}{2}, \frac{y_1 + y_2}{2}\right) = \left(\frac{}{2}, \frac{}{2}\right)$$

$$= \left(\frac{}{2}, \frac{}{2}\right) \text{ or } \boxed{}$$

The coordinates of the midpoint are $\boxed{}$.

EXAMPLE Find the Distance Between Two Points

2 What is the distance between $P(-1, 4)$ and $Q(2, -3)$?

$d = \sqrt{(x_2 - x_1)^2 + (y_2 - y_1)^2}$ Distance Formula

$= \sqrt{}$ Let $(x_1, y_1) = (-1, 4)$ and $(x_2, y_2) = (2, -3)$.

$= \sqrt{}$ Subtract.

$= \sqrt{} \text{ or } \sqrt{}$ Simplify.

The distance between the two points is $\boxed{}$ units.

Your Turn

a. Find the midpoint of the segment with endpoints at (3, 6) and (-1, -8).

b. What is the distance between $P(2, 3)$ and $Q(-3, 1)$?

Parabolas

© Glencoe/McGraw-Hill

WHAT YOU'LL LEARN

- Write equations of parabolas in standard form.

- Graph parabolas.

BUILD YOUR VOCABULARY (pages 178–179)

The graph of an equation of the form, $a \neq 0$,

[　　　　　　　] is a **parabola**.

Any figure that can be obtained by slicing a

[　　　　　] is called a **conic section**.

A parabola can also be defined as the set of all points in a

plane that are the same [　　　] from a given point

called the **focus** and a given line called the **directrix**.

EXAMPLE Analyze the Equation of a Parabola

1 Write $y = -x^2 - 2x + 3$ in standard form. Identify the vertex, axis of symmetry, and direction of opening of the parabola.

$y = -x^2 - 2x + 3$	Original equation
$y = -1(x^2 + 2x) + 3$	Factor.
$y = -1(x^2 + 2x + \square) + 3 - (-1)(\square)$	Complete the square.
$y = -1(x^2 + 2x + \boxed{}) + 3 + 1(\boxed{})$	Multiply $\boxed{}$ by -1.
$y = \boxed{}$	
$y = \boxed{}$	$(h, k) = (-1, 4)$

The vertex of this parabola is located at [　　　], and the

equation of the axis of symmetry is $x = \boxed{}$. The parabola opens downward.

KEY CONCEPT

Equation of a Parabola
The standard form of the equation of a parabola with vertex (h, k) and axis of symmetry $x = h$ is $y = a(x - h)^2 + k$.

- If $a > 0$, k is the minimum value of the related function and the parabola opens upward.

- If $a < 0$, k is the maximum value of the related function and the parabola opens downward.

Your Turn Write $y = 2x^2 + 4x + 5$ in standard form. Identify the vertex, axis of symmetry, and direction of opening of the parabola.

EXAMPLE Graph Parabolas

2 **a. Graph $y = 2x^2$.**

For this equation, $h = 0$ and $k = 0$. The vertex is at the origin. Substitute positive integers for x and find the corresponding y-values.

x	y
1	2
2	8
3	18

Since the graph is symmetric about

the y-axis, the points at [   ] ,

[   ] , and [   ] are also

on the parabola. Use all of these points to draw the graph.

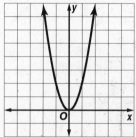

b. Graph $y = 2(x - 1)^2 - 5$.

The equation is of the form $y = a(x - h)^2 + k$, where $h = 1$ and $k = -5$. The graph of this equation is the graph of $y = 2x^2$

in part **a** translated [   ] unit

right and [   ] units down.

The vertex is now at [   ] .

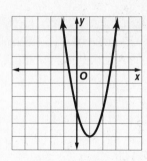

Your Turn Graph each equation.

a. $y = -3x^2$ **b.** $y = -3(x + 1)^2 + 3$

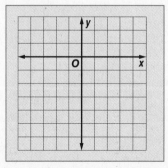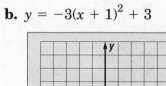

BUILD YOUR VOCABULARY (page 179)

The line segment through the focus of a parabola and

perpendicular to the axis of [   ] is called the

latus rectum.

EXAMPLE   **Graph an Equation Not in Standard Form**

③ Graph $x + y^2 = 4y - 1$.

First write the equation in the form $x = a(y - k)^2 + h$.

$x + y^2 = 4y - 1$

$x = -y^2 + 4y - 1$

$x = -1() - 1$

$x = -1(y^2 - 4y + \square) - 1 - (-1)(\square)$

$x = -1(y^2 - 4y + \boxed{}) - 1 + 1(\boxed{})$

$x = \boxed{}$

FOLDABLES

ORGANIZE IT

Use the tab for Vocabulary. Make a sketch of each vocabulary term in this lesson.

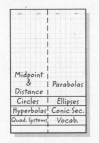

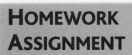

Then use the following information to draw the graph.

vertex: [   ]

axis of symmetry: $y =$ [   ]

focus: $\left(3 + \dfrac{1}{4(-1)}, 2\right)$ or $\left(\dfrac{11}{4}, 2\right)$

directrix: $x =$ [   ] or $3\dfrac{1}{4}$

direction of opening: left, since $a < 0$

length of the latus rectum: [   ] or [   ] unit

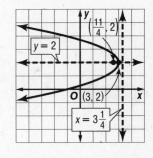

Your Turn   Graph $x - y^2 = 6y + 2$.

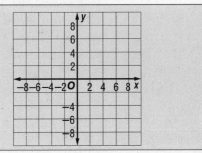

HOMEWORK ASSIGNMENT

Page(s):

Exercises:

WHAT YOU'LL LEARN

- Write equations of circles.
- Graph circles.

BUILD YOUR VOCABULARY (pages 178–179)

A circle is the set of all points in a plane that are

[ ] from a given [ ] in the plane,

called the **center**.

A line that [ ] the circle in exactly [ ]

point is said to be **tangent** to the circle.

EXAMPLE Write an Equation Given the Center and Radius

1 **LANDSCAPING** The plan for a park puts the center of a circular pond, of radius 0.6 miles, 2.5 miles east and 3.8 miles south of the park headquarters. Write an equation to represent the border of the pond, using the headquarters as the origin.

Since the headquarters is at [ ] , the center of the pond

is at [ ] with radius 0.6 mile.

KEY CONCEPT

Equation of a Circle
The equation of a circle with center (h, k) and radius r units is $(x - h)^2 + (y - k)^2 = r^2$

FOLDABLES On the tab for Circles, write this equation. Be sure to explain h, k, and r.

$(x - h)^2 + (y - k)^2 = r^2$ Equation of a circle

$(x - \boxed{})^2 + (y + \boxed{})^2 = \boxed{}^2$ $(h, k) = (2.5, -3.8)$, $r = 0.6$

$(x - 2.5)^2 + (y + 3.8)^2 = \boxed{}$ Simplify.

The equation is [ ]

Your Turn The plan for a park puts the center of a circular pond, of radius 0.5 mile, 3.5 miles west and 2.6 miles north of the park headquarters. Write an equation to represent the border of the pond, using the headquarters as the origin.

EXAMPLE Write an Equation Given a Diameter

2 **Write an equation for a circle if the endpoints of the diameter are at (2, 8) and (2, −2).**

First, find the center of the circle.

$(h, k) = \left(\dfrac{x_1 + x_2}{2}, \dfrac{y_1 + y_2}{2} \right)$ Midpoint Formula

$= \left(\dfrac{\boxed{}}{2}, \dfrac{\boxed{}}{2} \right)$ $(x_1, y_1) = (2, 8),$
$(x_2, y_2) = (2, -2)$

$= \left(\boxed{}, \boxed{} \right)$ Add.

$= \left(\boxed{} \right)$ Simplify.

Now find the radius.

$r = \sqrt{(x_2 - x_1)^2 + (y_2 - y_1)^2}$ Distance Formula

$= \sqrt{\boxed{}}$ $(x_1, y_1) = (2, 8),$
$(x_2, y_2) = (2, 3)$

$= \sqrt{\boxed{}}$ Subtract.

$= \boxed{}$ or $\boxed{}$ Simplify.

The radius of the circle is $\boxed{}$ units, so $r^2 = \boxed{}$.

Substitute h, k, and r^2 into the standard form of the equation of a circle. An equation of the circle is

.

Your Turn Write an equation for a circle if the endpoints of the diameter are at (3, 5) and (3, −7).

© Glencoe/McGraw-Hill

WRITE IT

What must you do to the equation of a circle if you want to graph the circle on a calculator?

_____

_____

_____

_____

EXAMPLE Graph an Equation in Standard Form

③ **Find the center and radius of the circle with equation $x^2 + y^2 = 16$. Then graph the circle.**

The center is at ☐ , and the

radius is ☐ .

The table lists some values for x and y that satisfy the equation.

Since the circle is centered at the origin, it is symmetric about the y-axis. Use these points and the concept of symmetry to graph $x^2 + y^2 = 16$.

x	y
0	4
1	3.9
2	3.5
3	2.6
4	0

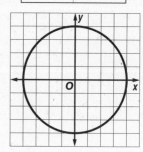

EXAMPLE Graph an Equation Not in Standard Form

④ **Find the center and radius of the circle with equation $x^2 + y^2 + 6x - 7 = 0$. Then graph the circle.**

Complete the square.

$$x^2 + 6x + \square + y^2 = 7$$

$$x^2 + 6x + \boxed{} + y^2 = 7 + \boxed{}$$

$$\boxed{} + y^2 = \boxed{}$$

The center is at ☐ , and the radius is ☐ .

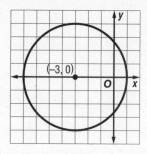

Your Turn Graph each equation.

a. $x^2 + y^2 = 9$

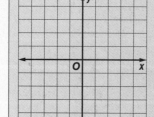

b. $x^2 + y^2 + 8x - 4y + 11 = 0$

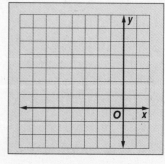

HOMEWORK ASSIGNMENT

Page(s):

Exercises:

_____

_____

Ellipses

What You'll Learn

- Write equations of ellipses.
- Graph ellipses.

BUILD YOUR VOCABULARY (pages 178–179)

An **ellipse** is the set of all points in a plane such that the

sum of the [_____] from two fixed [_____]

is constant. The two fixed points are called the **foci** of
the ellipse.

The points at which the ellipse intersects its axes of

symmetry determine two [_____] with

endpoints on the ellipse called the **major axis** and the
minor axis. The axes intersect at the **center** of the ellipse.

EXAMPLE Write an Equation for a Graph

1 **Write an equation for the ellipse shown.**

Write It

Write the name for the
endpoints of each axis
of an ellipse.

_____

_____

_____

_____

Find the values of a and b for the
ellipse. The length of the major axis of
any ellipse is $2a$ units. In this ellipse,
the length of the major axis is the
distance between $(0, 5)$ and $(0, -5)$.
10 units.

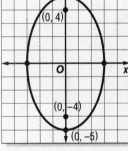

$2a = $ [____]　　Length of major axis = 10

$a = $ [____]　　Divide each side by 2.

The foci are located at $(0, 4)$ and $(0, -4)$, so $c = 4$.

Use the relationship between a, b, and c to find b.

$c^2 = a^2 - b^2$　　　　Equation relating a, b, and c

[____] $= $ [_____]　　$c = 4$ and $a = 5$

$b^2 = $ [____]　　　　Solve for b^2.

Since the major axis is vertical, substitute 25 for a^2 and 9 for
b^2 in the form $\dfrac{y^2}{a^2} + \dfrac{x^2}{b^2} = 1$. An equation of the ellipse is

[_____] .

Your Turn Write an equation for the ellipse shown.

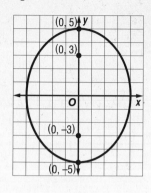

(0, 5)

(0, 3)

O   x

(0, −3)

(0, −5)

EXAMPLE **Write an Equation Given the Lengths of the Axes**

2 **SOUND** **A listener is standing in an elliptical room 150 feet wide and 320 feet long. When a speaker stands at one focus and whispers, the best place for the listener to stand is at the other focus. Write an equation to model this ellipse, assuming the major axis is horizontal and the center is at the origin.**

The length of the major axis is 320 feet.

$2a = 320$ Length of major axis = 320

$a = \boxed{}$ Divide each side by 2.

The length of the minor axis is 150 feet.

$2b = 150$ Length of minor axis = 150

$b = \boxed{}$ Divide each side by 2.

Substitute $a = \boxed{}$ and $b = \boxed{}$ into the form $\dfrac{x^2}{a^2} + \dfrac{y^2}{b^2} = 1$.

An equation for the ellipse is .

Your Turn A listener is standing in an elliptical room 60 feet wide and 120 feet long. When a speaker stands at one focus and whispers, the best place for the listener to stand is at the other focus. Write an equation to model this ellipse, assuming the major axis is horizontal and the center is at the origin.

EXAMPLE Graph an Equation Not in Standard Form

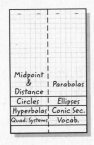

3 **Find the coordinates of the center and foci and the lengths of the major and minor axes of the ellipse with equation $x^2 + 4y^2 - 6x - 16y - 11 = 0$. Graph the ellipse.**

Complete the square to write in standard form.

$x^2 + 4y^2 - 6x - 16y - 11 = 0$

$x^2 - 6x + \square + 4(y^2 - 4y + \square) = 11 + \square + 4(\square)$

$(x^2 - 6x + \boxed{}) + 4(y^2 - 4y + \boxed{}) = 11 + 9 + 4(\boxed{})$

$$\boxed{}$$

$$\boxed{} = 1$$

The center is $\boxed{}$, and the foci are located at $\boxed{}$

and $\boxed{}$. The length of the major axis is

$\boxed{}$ units, and the length of the minor axis is $\boxed{}$.

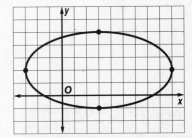

Your Turn Find the coordinates of the center and foci and the lengths of the major and minor axes of the ellipse with equation $4x^2 + 25y^2 + 16x - 150y + 141 = 0$. Graph the ellipse.

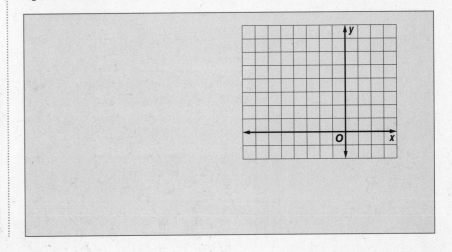

HOMEWORK ASSIGNMENT

Page(s):

Exercises:

WHAT YOU'LL LEARN

- Write equations of hyperbolas.
- Graph hyperbolas.

BUILD YOUR VOCABULARY (pages 178–179)

A **hyperbola** is the set of all points in a plane such that the [_____] value of the difference of the distances from two fixed points, called the **foci**, is constant.

As a hyperbola recedes from its center, the branches approach lines called **asymptotes**.

The **transverse axis** is a segment of length 2a whose endpoints are the vertices of the hyperbola.

The **conjugate axis** is a segment of length 2b units that is [_____] to the transverse axis at the center.

KEY CONCEPT

Equations of Hyperbolas with Centers at the Origin

Horizontal Transverse Axis $\dfrac{x^2}{a^2} - \dfrac{y^2}{b^2} = 1$

Vertical Transverse Axis $\dfrac{y^2}{a^2} - \dfrac{x^2}{b^2} = 1$

EXAMPLE Write an Equation for a Graph

1 Write an equation for the hyperbola.

The center is [_____].

The value of a is the distance from the center to a vertex, or [___] units. The value of c is the distance from the center to a focus, or [___] units.

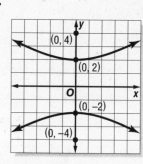

$c^2 = a^2 + b^2$ Equation relating a, b, and c for a hyperbola

[___] = [___] $c = 4$, $a = 2$

[___] = [___]   Evaluate the squares.

$12 = b^2$ Solve for b^2.

Since the transverse axis is vertical, the equation is of the form $\dfrac{y^2}{a^2} - \dfrac{x^2}{b^2} = 1$. Substitute the values for a^2 and b^2.

An equation of the hyperbola is 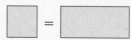.

Your Turn Write an equation for the hyperbola.

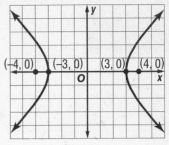

(–4, 0) (–3, 0) (3, 0) (4, 0)

EXAMPLE Graph an Equation Not in Standard Form

2 **Find the coordinates of the vertices and foci and the equations of the asymptotes for the hyperbola with equation $x^2 - y^2 + 6x + 10y - 17 = 0$. Then graph the hyperbola.**

Complete the square for each variable.

$$x^2 - y^2 + 6x + 10y - 17 = 0$$

$$x^2 - 6x + \square - 1(y^2 - 10y + \square) = 17 + \square - 1(\square)$$

$$x^2 + 6x + \boxed{} - 1(y^2 - 10y + \boxed{}) = 17 + \boxed{} - 1(25)$$

$$\boxed{} - \boxed{} = \boxed{}$$

The vertices are $\boxed{}$ and

$\boxed{}$, and the foci are

$(\sqrt{2} - 3, 5)$ and $(-\sqrt{2} - 3, 5)$.

The equations of the asymptotes are

$\boxed{}$ or $y = \boxed{}$ and $y = -x + 2$.

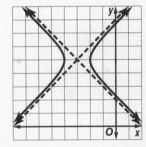

Your Turn Find the coordinates of the vertices and foci and the equations of the asymptotes for the hyperbola with equation $9x^2 - 16y^2 - 72x - 64y + 224 = 0$. Then graph the hyperbola.

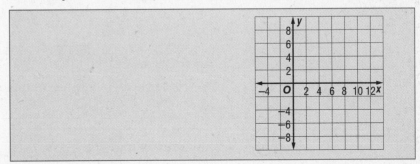

FOLDABLES

ORGANIZE IT

Under the tab for Hyperbolas, describe two similarities between hyperbolas and ellipses.

Midpoint & Distance | Parabolas
Circles | Ellipses
Hyperbolas | Conic Sec.
Quad. Systems | Vocab.

HOMEWORK ASSIGNMENT

Page(s):

Exercises:

Conic Sections

WHAT YOU'LL LEARN

- Write equations of conic sections in standard form
- Identify conic sections from their equations.

REMEMBER IT

If $a = b$ in the equation for an ellipse, the graph of the equation is a circle.

EXAMPLE Rewrite an Equation of a Conic Section

1 Write the equation $y^2 = 18 - 2x^2$ in standard form. State whether the graph of the equation is a *parabola, circle, ellipse,* or *hyperbola.* Then graph the equation.

Write the equation in standard form.

$$y^2 = 18 - 2x^2$$

$$\boxed{} = \boxed{} \quad \text{Isolate terms}$$

$$\boxed{} = \boxed{} \quad \text{Divide each side by } \boxed{}.$$

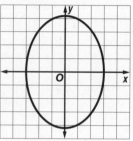

The graph is an ellipse with center at $\boxed{}$.

Your Turn Write the equation $x^2 + y^2 - 6x - 7 = 0$ in standard form. State whether the graph of the equation is a parabola, circle, ellipse, or hyperbola. Then graph the equation.

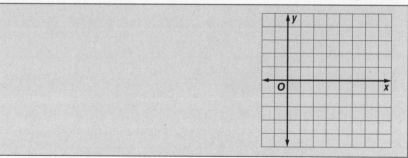

EXAMPLE Analyze an Equation of a Conic Section

2 Without writing the equation in standard form, state whether the graph of the equation is a *parabola, circle, ellipse,* or *hyperbola.*

a. $3y^2 - x^2 - 9 = 0$

$A = \boxed{}$ and $C = \boxed{}$

Since $\boxed{}$,

the graph is a $\boxed{}$.

ORGANIZE IT

Under the tab for Conic Sections, sketch and label each of the conic sections. Then write the standard form on the conic section below each label.

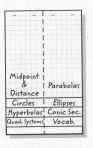

Midpoint & Distance	Parabolas
Circles	Ellipses
Hyperbolas	Conic Sec.
Quad. Systems	Vocab.

b. $2x^2 + 2y^2 + 16x - 20y = -32$

$A = \boxed{}$ and $C = \boxed{}$

Since $\boxed{}$, the graph is a $\boxed{}$.

c. $y^2 - 2x - 4y + 10 = 0$

$A = \boxed{}$ and $C = \boxed{}$

Since $\boxed{}$, this graph is a $\boxed{}$.

Your Turn Without writing the equation in standard form, state whether the graph of the equation is a *parabola, circle, ellipse,* or *hyperbola.*

a. $2y^2 - x^2 + 16 = 0$

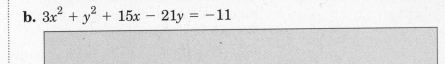

b. $3x^2 + y^2 + 15x - 21y = -11$

c. $y^2 - 3x + 2y - 10 = 0$

HOMEWORK ASSIGNMENT

Page(s):

Exercises:

Solving Quadratic Systems

WHAT YOU'LL LEARN

- Solve systems of quadratic equations algebraically and graphically.

- Solve systems of quadratic inequalities graphically.

EXAMPLE Linear-Quadratic System

1 **Solve the system of equations.**

$$4x^2 - 16y^2 = 25$$
$$2y + x = 2$$

You can use a graphing calculator to help visualize the relationships of the graphs of the equations and predict the number of solutions.

Solve each equation for y to obtain

$$y = \pm \frac{\sqrt{4x^2 - 25}}{4} \text{ and}$$

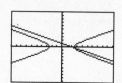

$$y = \boxed{}.$$

REVIEW IT

By the Square Root Property, for any real number n, if $x^2 = n$, then $x = $ _____ ?
(Lesson 6-4)

Enter the functions on the Y= screen. The graph indicates that the hyperbola and the line intersect in one point. So, the system has one solution.

Use substitution to solve the system.

First, rewrite $2y + x = 2$ as $x = 2 - 2y$.

$$4x^2 - 16y^2 = 25 \qquad \text{First equation in the system}$$

$$4(\boxed{})^2 - 16y^2 = 25 \qquad \text{Substitute } \boxed{} \text{ for } x.$$

$$\boxed{} + 16 = 25 \qquad \text{Simplify.}$$

$$\boxed{} = \boxed{} \qquad \text{Subtract } \boxed{} \text{ from each side.}$$

$$y = \boxed{} \qquad \text{Divide each side by } \boxed{}.$$

Now solve for x.

$$x = 2 - 2y \qquad \text{Equation for } x \text{ in terms of } y$$

$$x = 2 - 2\left(\boxed{}\right) \qquad \text{Substitute the } y\text{-value.}$$

$$x = \boxed{} \qquad \text{Simplify.}$$

The solution is $\left(\boxed{}\right)$.

Your Turn Solve $x^2 - y^2 = 4$ and $2y + x = 2$.

EXAMPLE Quadratic-Quadratic System

2 **Solve the system of equations.**

$x^2 + y^2 = 16$
$4x^2 + y^2 = 23$

A graphing calculator indicates that the circle and ellipse intersect in four points. So, this system has four solutions.

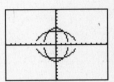

Use the elimination method to solve the system.

$-x^2 - y^2 = -16$ Rewrite the first original equation.

$\underline{(+)\ 4x^2 + y^2 = 23}$ Second original equation

$\boxed{} = 7$ Add.

$\boxed{} = \boxed{}$ Divide each side by 3.

$\boxed{} = \pm\,\boxed{}$ Take the square root of each side.

Substitute $\sqrt{\dfrac{7}{3}}$ and $-\sqrt{\dfrac{7}{3}}$ 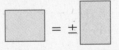 in either of the original equations and solve for y.

$x^2 + y^2 = 16$	$x^2 + y^2 = 16$	Original equation
$\left(\sqrt{\dfrac{7}{3}}\right)^2 + y^2 = 16$	$\left(-\sqrt{\dfrac{7}{3}}\right)^2 + y^2 = 16$	Substitute for x.
$y^2 = \boxed{}$	$y^2 = \boxed{}$	Subtract from each side.
$y = \boxed{}$	$y = \boxed{}$	Take the square root of each side.

The solutions are $\boxed{}$, $\boxed{}$,

$\boxed{}$  , and $\boxed{}$.

REMEMBER IT

When graphing conic sections, press ZOOM 5. This window gives the graphs a more realistic look.

Your Turn Solve $x^2 + y^2 = 10$ and $3x^2 + y^2 = 28$.

EXAMPLE System of Quadratic Inequalities

3 **Solve the system of inequalities by graphing.**

$y > x^2 + 1$
$x^2 + y^2 \leq 9$

The graph of $y > x^2 + 1$ is the

[   ] $y = x^2 + 1$ and the

region inside and above it. Shade
the region dark gray.

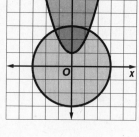

The graph of $x^2 + y^2 \leq 9$ is the

interior of [   ] $x^2 + y^2 = 9$.

Shade the region medium gray.

The intersection of these regions, shaded darker gray,
represents the solution of the system of inequalities.

Your Turn Solve $y < -x^2 + 1$ and $x^2 + y^2 \leq 4$ by graphing.

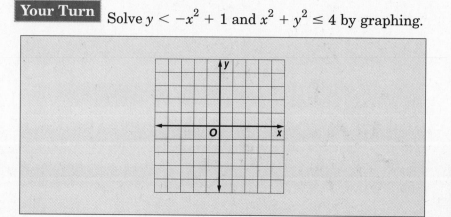

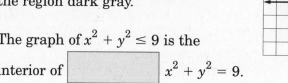

FOLDABLES

ORGANIZE IT

Under the tab of
Quadratic Systems,
sketch five graphs of
systems of quadratic
equations. Write the
number of solutions
below each graph.

Midpoint & Distance	Parabolas
Circles	Ellipses
Hyperbolas	Conic Sec.
Quad. Systems	Vocab.

HOMEWORK
ASSIGNMENT

Page(s):

Exercises:

BRINGING IT ALL TOGETHER

STUDY GUIDE

FOLDABLES™	VOCABULARY PUZZLEMAKER	**BUILD YOUR VOCABULARY**
Use your **Chapter 8 Foldable** to help you study for your chapter test.	To make a crossword puzzle, word search, or jumble puzzle of the vocabulary words in Chapter 8, go to: www.glencoe.com/sec/math/ t_resources/free/index.php	You can use your completed **Vocabulary Builder** (pages 178–179) to help you solve the puzzle.

8-1
Midpoint and Distance Formulas

Consider the segment connecting the points $(-3, 5)$ and $(9, 11)$.

1. Find the midpoint of this segment.

2. Find the length of the segment. Write your answer in simplified radical form.

3. Circle P has a diameter $\overline{CD}$. If C is at $(4, -3)$ and D is at $(-3, 5)$, find the center of the circle and the length of the diameter.

8-2
Parabolas

Write each equation in standard form.

4. $y = 2x^2 - 8x + 1$

5. $y = -2x^2 + 6x + 1$

6. $y = \frac{1}{2}x^2 - 5x + 12$

8-3
Circles

7. Write the equation of the circle with center $(4, -3)$ and radius 5.

[   ]

8. The circle with equation $(x + 8)^2 + y^2 = 121$ has center

[   ] and radius [   ].

9. a. In order to find the center and radius of the circle with equation

$x^2 + y^2 + 4x - 6y - 3 = 0$, it is necessary to [   ].

Fill in the missing parts of this process.

$$x^2 + y^2 + 4x - 6y - 3 = 0$$

$$x^2 + y^2 + 4x - 6y = \boxed{}$$

$$x^2 + 4x + \boxed{} + y^2 - 6y + \boxed{} = \boxed{} + \boxed{} + \boxed{}$$

$$(x + \boxed{})^2 + (y - \boxed{})^2 = \boxed{}$$

b. This circle has radius 4 and center at [   ].

8-4
Ellipses

10.

Standard Form of Equation	$\frac{y^2}{25} + \frac{x^2}{16} = 1$	$\frac{x^2}{9} + \frac{y^2}{4} = 1$
Direction of Major Axis		
Direction of Minor Axis		
Foci		
Length of Major Axis		
Length of Minor Axis		

Complete the table to describe characteristics of their graphs.

8-5
Hyperbolas

Study the hyperbola graphed at the right.

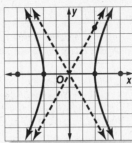

11. The center is ⬚.

12. The value of a is ⬚.

13. The value of c is ⬚.

14. To find b^2, solve ⬚ = ⬚ + ⬚.

15. The equation in standard form for this hyperbola is ⬚.

8-6
Conic Sections

Name the conic section that is the graph of each equation. Give the coordinates of the vertex if the conic section is a parabola and of the center if it is a circle, an ellipse, or a hyperbola.

16. $\dfrac{(x-3)^2}{36} + \dfrac{(y+5)^2}{15} = 1$

17. $x = -2(y+1)^2 + 7$

18. $(x-5)^2 - (y+5)^2 = 1$

19. $(x+6)^2 + (y-2)^2 = 1$

8-7
Solving Quadratic Systems

Draw a sketch to illustrate each of the following possibilities.

20. a parabola and a line that intersect in 2 points

21. an ellipse and a circle that intersect in 4 points

22. a hyperbola and a line that intersect in 1 point

ARE YOU READY FOR THE CHAPTER TEST?

Math Online

Visit **algebra2.com** to access your textbook, more examples, self-check quizzes, and practice tests to help you study the concepts in Chapter 8.

Check the one that applies. Suggestions to help you study are given with each item.

☐ **I completed the review of all or most lessons without using my notes or asking for help.**

- You are probably ready for the Chapter Test.
- You may want to take the Chapter 8 Practice Test on page 467 of your textbook as a final check.

☐ **I used my Foldable or Study Notebook to complete the review of all or most lessons.**

- You should complete the Chapter 8 Study Guide and Review on pages 461–466 of your textbook.
- If you are unsure of any concepts or skills, refer back to the specific lesson(s).
- You may also want to take the Chapter 8 Practice Test on page 467 of your textbook.

☐ **I asked for help from someone else to complete the review of all or most lessons.**

- You should review the examples and concepts in your Study Notebook and Chapter 8 Foldable.
- Then complete the Chapter 8 Study Guide and Review on pages 461–466 of your textbook.
- If you are unsure of any concepts or skills, refer back to the specific lesson(s).
- You may also want to take the Chapter 8 Practice Test on page 467 of your textbook.

Student Signature	Parent/Guardian Signature

Teacher Signature

CHAPTER 9

Rational Expressions and Equations

 Use the instructions below to make a Foldable to help you organize your notes as you study the chapter. You will see Foldable reminders in the margin of this Interactive Study Notebook to help you in taking notes.

Begin with a sheet of plain $8\frac{1}{2}$" × 11" paper.

STEP 1 **Fold**
Fold in half lengthwise leaving a $1\frac{1}{2}$" margin at the top. Fold again in thirds.

STEP 2 **Cut and Label**
Open. Cut along the second folds to make three tabs. Label as shown.

 NOTE-TAKING TIP: When you take notes, it may be helpful to create a concept map by writing the main idea or key words.

CHAPTER
9

This is an alphabetical list of new vocabulary terms you will learn in Chapter 9. As you complete the study notes for the chapter, you will see Build Your Vocabulary reminders to complete each term's definition or description on these pages. Remember to add the textbook page number in the second column for reference when you study.

Vocabulary Term	Found on Page	Definition	Description or Example
asymptote [A-suhm(p)-TOHT]			
complex fraction			
constant of variation			
continuity [KAHN-tuhn-OO-uh-tee]			
direct variation			
inverse variation [IHN-VUHRS]			
joint variation			
point discontinuity			
rational equation			
rational expression			
rational function			
rational inequality			

Multiplying and Dividing Rational Expressions

WHAT YOU'LL LEARN

- Simplify rational expressions.
- Simplify complex fractions.

BUILD YOUR VOCABULARY (page 202)

A ratio of two [____] expressions is called a **rational expression**.

EXAMPLE Simplify a Rational Expression

1 **a.** Simplify $\dfrac{3y(y+7)}{(y+7)(y^2-9)}$.

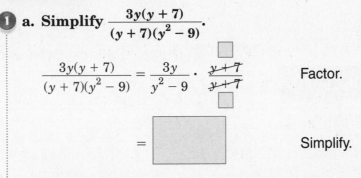

$$\frac{3y(y+7)}{(y+7)(y^2-9)} = \frac{3y}{y^2-9} \cdot \frac{\cancel{y+7}}{\cancel{y+7}}$$ Factor.

$$= \boxed{}$$ Simplify.

b. Under what conditions is this expression undefined?

To find out when this expression is undefined, completely factor the denominator.

$$\frac{3y(y+7)}{(y+7)(y^2-9)} = \frac{3y(y+7)}{(y+\,)\boxed{}}$$

The values that would make the denominator equal to 0 are [____] , [____] , and [____] .

Your Turn

a. Simplify $\dfrac{x(x+5)}{(x+5)(x^2-16)}$.

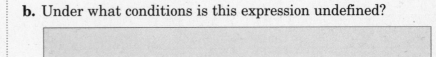

b. Under what conditions is this expression undefined?

EXAMPLE Simplify by Factoring Out −1

② Simplify $\dfrac{a^4b - 2a^4}{2a^3 - a^3b}$.

$$\dfrac{a^4b - 2a^4}{2a^3 - a^3b} = \boxed{}$$

Factor the numerator and the denominator.

$$= \dfrac{\overset{a}{\cancel{a^4}}(-1)(2 \cancel{- b})}{\underset{1}{\cancel{a^3}}\underset{1}{(2 \cancel{- b})}}$$

$b - 2 = -(-b + 2)$ or $-1(2 - b)$

$$= \boxed{} \text{ or } -a$$

Simplify.

EXAMPLE Multiply Rational Expressions

KEY CONCEPT

Rational Expressions

- To multiply two rational expressions, multiply the numerators and the denominators.

- To divide two rational expressions, multiply by the reciprocal of the divisor.

③ Simplify $\dfrac{8x}{21y^3} \cdot \dfrac{7y^2}{16x^3}$.

$$\dfrac{8x}{21y^3} \cdot \dfrac{7y^2}{16x^3} = \dfrac{\overset{1}{\cancel{2}} \cdot \overset{1}{\cancel{2}} \cdot \overset{1}{\cancel{2}} \cdot \overset{1}{\cancel{x}} \cdot \overset{1}{\cancel{7}} \cdot \overset{1}{\cancel{y}} \cdot \overset{1}{\cancel{y}}}{3 \cdot \underset{1}{\cancel{7}} \cdot \underset{1}{\cancel{y}} \cdot \underset{1}{\cancel{y}} \cdot y \cdot \underset{1}{\cancel{2}} \cdot \underset{1}{\cancel{2}} \cdot \underset{1}{\cancel{2}} \cdot 2 \cdot \underset{1}{\cancel{x}} \cdot x \cdot x}$$

Factor.

$$= \boxed{} \text{ or } \boxed{}$$

Simplify.

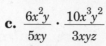 **Your Turn** Simplify each expression.

a. $\dfrac{x^4y - 3x^4}{3x^3 - x^3y}$

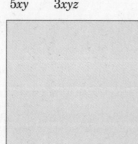

b. $\dfrac{3x}{15y} \cdot \dfrac{5y^2}{2x^3}$

c. $\dfrac{6x^2y}{5xy} \cdot \dfrac{10x^3y^2}{3xyz}$

d. $\dfrac{3x^2y}{20ab} \div \dfrac{6xy}{5a^2b^3}$

EXAMPLE Polynomials in the Numerator and Denominator

4 Simplify $\dfrac{k-3}{k+1} \div \dfrac{k^2-4k+3}{1-k^2}$.

$\dfrac{k-3}{k+1} \div \dfrac{k^2-4k+3}{1-k^2}$

$= \dfrac{k-3}{k+1} \cdot$ 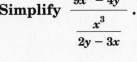 Multiply by the reciprocal of the divisor.

$= \dfrac{(k-3)(1-k)(1+k)}{(k+1)(k-1)(k-3)}$ $1+k=k+1,$
$1-k=-1(k-1)$

$=$ 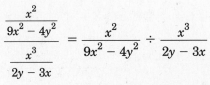 Simplify.

EXAMPLE Simplify a Complex Fraction

5 Simplify $\dfrac{\dfrac{x^2}{9x^2-4y^2}}{\dfrac{x^3}{2y-3x}}$.

$\dfrac{\dfrac{x^2}{9x^2-4y^2}}{\dfrac{x^3}{2y-3x}} = \dfrac{x^2}{9x^2-4y^2} \div \dfrac{x^3}{2y-3x}$ Express as a division expression.

$= \dfrac{x^2}{9x^2-4y^2} \cdot$ 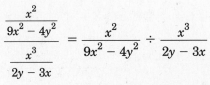 Multiply by the reciprocal of divisor.

$= \dfrac{x \cdot x(2y-3x)}{(3x-2y)(3x+2y)\,x \cdot x \cdot x}$ Factor.

$=$ Simplify.

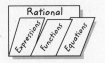

ORGANIZE IT
Under the tab for Rational Expressions, write how simplifying rational expressions is similar to simplifying rational numbers.

HOMEWORK ASSIGNMENT
Page(s):
Exercises:

Your Turn Simplify each expression.

a. $\dfrac{x-3}{x+2} \cdot \dfrac{x^2+5x+6}{x^2-9}$

b. $\dfrac{\dfrac{a^2}{a^2-9b^2}}{\dfrac{a^4}{a+3b}}$

Adding and Subtracting Rational Expressions

EXAMPLE LCM of Monomials

WHAT YOU'LL LEARN

- Determine the LCM of polynomials.

- Add and subtract rational expressions.

1 Find the LCM of $15a^2bc^3$, $16b^5c^2$, and $20a^3c^6$.

$15a^2bc^3 =$

Factor the first monomial.

$16b^5c^2 =$

Factor the second monomial.

$20a^3c^6 =$

Factor the third monomial.

LCM =

$=$

Use each factor the greatest number of times it appears as a factor and simplify.

EXAMPLE LCM of Polynomials

2 Find the LCM of $x^3 - x^2 - 2x$ and $x^2 - 4x + 4$.

$x^3 - x^2 - 2x =$

Factor the first polynomial.

$x^2 - 4x + 4 =$

Factor the second polynomial.

LCM =

Use each factor the greatest number of times it appears as a factor.

Your Turn Find the LCM of each set of polynomials.

a. $6x^2zy^3$, $9x^3y^2z^2$, $4x^2z$

b. $x^3 + 2x^2 - 3x$, $x^2 + 6x + 9$

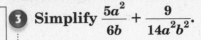

EXAMPLE Monomial Denominators

3 Simplify $\dfrac{5a^2}{6b} + \dfrac{9}{14a^2b^2}$.

$$\dfrac{5a^2}{6b} + \dfrac{9}{14a^2b^2}$$

$$= \dfrac{5a^2 \cdot \boxed{}}{6b \cdot \boxed{}} + \dfrac{9 \cdot \boxed{}}{14a^2b^2 \cdot \boxed{}}$$

The LCD is $42a^2b^2$. Find equivalent fractions that have this denominator.

$$= \dfrac{\boxed{}}{42a^2b^2} + \dfrac{\boxed{}}{42a^2b^2}$$

Simplify each numerator and denominator.

$$= \dfrac{35a^4b + 27}{42a^2b^2}$$

Add the numerators.

EXAMPLE Polynomial Denominators

4 Simplify $\dfrac{x + 10}{3x - 15} - \dfrac{3x + 15}{6x - 30}$.

$$\dfrac{x + 10}{3x - 15} - \dfrac{3x + 15}{6x - 30}$$

$$= \dfrac{x + 10}{\boxed{}} - \dfrac{3x + 15}{\boxed{}}$$

Factor the denominators.

$$= \dfrac{2(x + 10)}{2 \cdot 3(x - 5)} - \dfrac{3x + 15}{6(x - 5)}$$

The LCD is $6(x - 5)$.

$$= \dfrac{2(x + 10) - (3x + 15)}{6(x - 5)}$$

Subtract the numerators.

$$= \dfrac{\boxed{}}{6(x - 5)}$$

Distributive Property

$$= \dfrac{\boxed{}}{6(x - 5)}$$

Combine like terms.

$$= \dfrac{-1(\cancel{x - 5})^{1}}{6(\cancel{x - 5})_{1}} \text{ or } \boxed{}$$

Simplify.

Your Turn Simplify each expression.

a. $\dfrac{3x^2}{2y} + \dfrac{5}{12xy^2}$

b. $\dfrac{x+5}{2x-4} - \dfrac{3x+8}{4x-8}$

EXAMPLE Simplify Complex Fractions

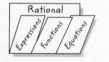

ORGANIZE IT

Under the tab for Rational Expressions, write your own example similar to Example 5. Then simplify your rational expression. Give the reason for each step.

❺ Simplify $\dfrac{\dfrac{1}{a} + \dfrac{1}{b}}{\dfrac{1}{b} - 1}$.

$$\dfrac{\dfrac{1}{a} + \dfrac{1}{b}}{\dfrac{1}{b} - 1} = \dfrac{\dfrac{b}{ab} + \dfrac{a}{ab}}{\dfrac{1}{b} - \dfrac{b}{b}}$$

The LCD of the numerator is ab.
The LCD of the denominator is b.

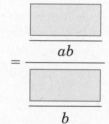

$$= \dfrac{\boxed{}}{ab}$$
$$\dfrac{\boxed{}}{b}$$

Simplify the numerator and denominator.

$$= \boxed{} \div \boxed{}$$

Write as a division expression.

$$= \dfrac{b+a}{a\cancel{b}_1} \cdot \dfrac{\cancel{b}^1}{1-b}$$

Multiply by the reciprocal of the divisor.

$$= \dfrac{b+a}{a(1-b)} \text{ or } \boxed{}$$

Simplify.

Your Turn Simplify $\dfrac{\dfrac{1}{a} - \dfrac{1}{b}}{\dfrac{1}{b} + \dfrac{1}{a}}$.

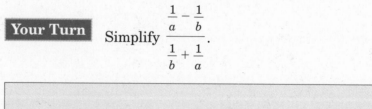

HOMEWORK ASSIGNMENT

Page(s):

Exercises:

Graphing Rational Functions

WHAT YOU'LL LEARN

- Determine the vertical asymptotes and the point discontinuity for the graphs of rational functions.

- Graph rational functions.

BUILD YOUR VOCABULARY (page 202)

A **rational function** is an equation of the form

$f(x) = \dfrac{p(x)}{q(x)}$, where $p(x)$ and $q(x)$ are [   ]

functions and $q(x) \neq 0$.

The graphs of rational functions may have breaks in **continuity**. This means that not all rational functions are

traceable. Breaks in [   ] can appear as a

vertical **asymptote** or as a **point discontinuity**.

EXAMPLE Vertical Asymptotes and Point Discontinuity

KEY CONCEPTS

Vertical Asymptotes
If the rational expression of a function is written in simplest form and the function is undefined for $x = a$, then $x = a$ is a vertical asymptote.

Point Discontinuity
If the original function is undefined for $x = a$ but the rational expression of the function in simplest form is defined for $x = a$, then there is a hole in the graph at $x = a$.

① Determine the equations of any vertical asymptotes and the values of x for any holes in the graph of

$$f(x) = \frac{x^2 - 4}{x^2 + 5x + 6}.$$

Factor the numerator and denominator of the rational expression.

$$\frac{x^2 - 4}{x^2 + 5x + 6} = \boxed{}$$

The function is undefined for $x = \boxed{}$ and $\boxed{}$.

Since $\dfrac{(x - 2)\overset{1}{\cancel{(x + 2)}}}{\underset{1}{\cancel{(x + 2)}}(x + 3)} = \boxed{}$, $x = \boxed{}$ is a vertical

asymptote and $x = \boxed{}$ is a hole in the graph.

Your Turn   Determine the equations of any vertical asymptotes and the values of x for any holes in the graph

of $f(x) = \dfrac{x^2 - 9}{x^2 + 8x + 15}$.

ORGANIZE IT

Under the tab for Rational Functions, write how you can tell from a rational function where the breaks in continuity will appear in the graph of the function.

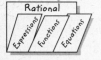

EXAMPLE Graph with a Vertical Asymptote

2 Graph $f(x) = \dfrac{x}{x+1}$.

The function is undefined for $x = $ ☐ . Since $\dfrac{x}{x+1}$ is in its simplest form, $x = $ ☐ is a vertical asymptote. Make a table of values. Plot the points and draw the graph.

x	$f(x)$
-4	1.33
-3	1.5
-2	2
0	0
1	0.5
2	0.67
3	0.75

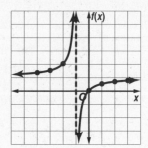

EXAMPLE Graph with Point Discontinuity

3 Graph $f(x) = \dfrac{x^2 - 4}{x - 2}$.

Notice that $\dfrac{x^2 - 4}{x - 2} = \dfrac{(x+2)(x-2)}{x - 2}$

or ☐ . Therefore, the

graph of $f(x) = \dfrac{x^2 - 4}{x - 2}$ is the graph of

$f(x) = $ ☐ with a hole at $x = $ ☐ .

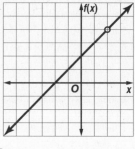

Your Turn Graph each rational function.

a. $f(x) = \dfrac{x}{x+3}$

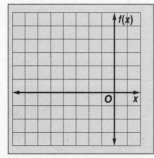

b. $f(x) = \dfrac{x^2 - 16}{x + 4}$

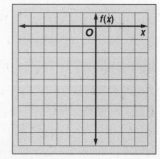

HOMEWORK ASSIGNMENT

Page(s):

Exercises:

Direct, Joint, and Inverse Variation

WHAT YOU'LL LEARN

- Recognize and solve direct and joint variation problems.

- Recognize and solve inverse variation problems.

KEY CONCEPTS

Direct Variation y varies directly as x if there is some nonzero constant k such that $y = kx$. k is called the constant of variation.

Joint Variation y varies jointly as x and z if there is some number k such that $y = kxz$, where $k \neq 0$, $x \neq 0$, and $z \neq 0$.

EXAMPLE Direct Variation

1 **If y varies directly as x and $y = -15$ when $x = 5$, find y when $x = 3$.**

Use a proportion that relates the values.

$$\frac{y_1}{x_1} = \frac{y_2}{x_2}$$ Direct proportion.

$$\boxed{} = \boxed{}$$ $y_1 = -15$, $x_1 = 5$, and $x_2 = 3$

$$\boxed{} = \boxed{}$$ Cross multiply.

$$\boxed{} = \boxed{}$$ Simplify.

$$\boxed{} = y_2$$ Divide each side by 5.

When $x = 3$, the value of y is $\boxed{}$.

EXAMPLE Joint Variation

2 **Suppose y varies jointly as x and z. Find y when $x = 10$ and $z = 5$ if $y = 12$ when $x = 3$ and $z = 8$.**

Use a proportion that relates the values.

$$\frac{y_1}{x_1 z_1} = \frac{y_2}{x_2 z_2}$$ Joint variation.

$$\boxed{} = \boxed{}$$ $y_1 = 12$, $x_1 = 3$, $z_1 = 8$, $x_2 = 10$, and $z_2 = 5$

$$\boxed{} = \boxed{}$$ Cross multiply.

$$\boxed{} = \boxed{}$$ Simplify.

$$\boxed{} = y_2$$ Divide each side by 24.

When $x = 10$ and $z = \boxed{}$, $y = \boxed{}$.

EXAMPLE Inverse Variation

KEY CONCEPT

Inverse Variation y varies inversely as x if there is some nonzero constant k such that $xy = k$ or $y = \dfrac{k}{x}$.

③ **If a varies inversely as b and $a = -6$ when $b = 2$, find a when $b = -7$.**

Use a proportion that relates the values.

$$\frac{a_1}{b_2} = \frac{a_2}{b_1}$$ Inverse variation

$$\boxed{} = \boxed{}$$ $a_1 = -6$, $b_1 = 2$, and $b_2 = -7$

$$\boxed{} = \boxed{}$$ Cross multiply.

$$\boxed{} = \boxed{}$$ Simplify.

$$\boxed{} = \boxed{}$$ Divide each side by -7.

When $b = -7$, a is $\boxed{}$ or $1\frac{5}{7}$.

Your Turn

a. If y varies directly as x and $y = 12$ when $x = -3$, find y when $x = 7$.

b. Suppose y varies jointly as x and z. Find y when $x = 3$ and $z = 2$ if $y = 11$ when $x = 5$ and $z = 22$.

c. If a varies inversely as b and $a = 3$ when $b = 8$, find a when $b = 6$.

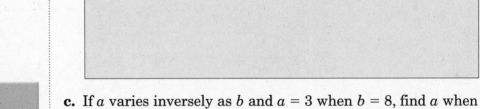

HOMEWORK ASSIGNMENT

Page(s):

Exercises:

Classes of Functions

EXAMPLE Identify a Function Given the Graph

1 Identify the type of function represented by each graph.

a.

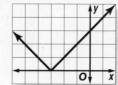

The graph is a V shape, so it is

b.

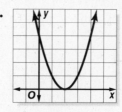

The graph is a parabola, so it is

Your Turn Identify the type of function represented by each graph.

a.

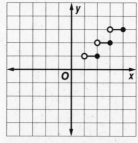

b.

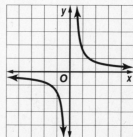

EXAMPLE Identify a Function Given its Equation

2 **Identify the type of function represented by each equation. Then graph the equation.**

a. $y = -3$

Since the equation has no x-intercept, it is the constant function. Determine some points on the graph and graph it.

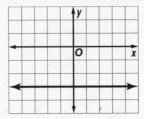

b. $y = \sqrt{9x}$

Since the equation includes an expression with a square root, it is a square root function. Plot some points and use what you know about square root graphs to graph it.

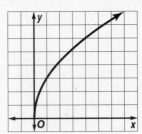

Your Turn **Identify the type of function represented by each equation. Then graph the equation.**

a. $y = \dfrac{x^2 - 9}{x + 3}$

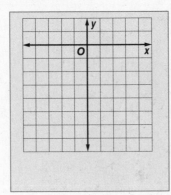

b. $y = -2x$

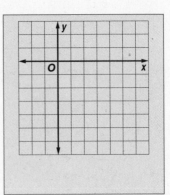

Solving Rational Equations and Inequalities

WHAT YOU'LL LEARN

- Solve rational equations.
- Solve rational inequalities.

REMEMBER IT

Rational equations are often easier to solve when you eliminate the fractions first.

BUILD YOUR VOCABULARY (page 202)

Any equation that contains one or more rational expressions is called a **rational equation**.

Inequalities that contain one or more rational expressions are called **rational inequalities**.

EXAMPLE Solve a Rational Equation

1 Solve $\dfrac{5}{24} + \dfrac{2}{3-x} = \dfrac{1}{4}$.

The LCD for the three denominators is $24(3 - x)$.

$$\frac{5}{24} + \frac{2}{3-x} = \frac{1}{4} \qquad \text{Original equation}$$

$$\boxed{}\left(\frac{5}{24} + \frac{2}{3-x}\right) = \frac{1}{4}\boxed{} \qquad \text{Multiply each side by } \boxed{}.$$

$$\overset{1}{\cancel{24}}(3-x)\left(\frac{5}{\cancel{24}_1}\right) + 24\overset{1}{(\cancel{3-x})}\left(\frac{2}{\cancel{3-x}_1}\right) = \frac{1}{\underset{1}{\cancel{4}}}\overset{6}{\cancel{24}}(3-x) \qquad \text{Simplify.}$$

$$\boxed{} = \boxed{} \qquad \text{Simplify.}$$

$$\boxed{} = \boxed{} \qquad \text{Add.}$$

$$x = \boxed{}$$

Your Turn Solve $\dfrac{5}{2} + \dfrac{3}{x-1} = \dfrac{1}{2}$.

EXAMPLE Elimination of a Possible Solution

2 Solve $\dfrac{p^2 - p + 1}{p + 1} = \dfrac{p^2 - 7}{p^2 - 1} + p.$

The LCD is $p^2 - 1$.

$$\dfrac{p^2 - p + 1}{p + 1} = \dfrac{p^2 - 7}{p^2 - 1} + p \qquad \text{Original equation}$$

$$\overset{p-1}{(p^2 - 1)}\dfrac{p^2 - p + 1}{\underset{1}{p+1}} = \dfrac{p^2 - 7}{\underset{1}{p^2-1}}\overset{1}{(p^2-1)} + p(p^2 - 1)$$

$$(p - 1)(p^2 - p + 1) = p^2 - 7 + (p^2 - 1)p \qquad \text{Distributive Property}$$

$$p^3 - p^2 + p - p^2 + p - 1 = p^2 - 7 + p^3 - p \qquad \text{Simplify.}$$

$$-2p^2 + 2p - 1 = p^2 - p - 7 \qquad \text{Simplify.}$$

$$\boxed{} = \boxed{} \qquad \begin{array}{l}\text{Add} \\ (2p^2 - 2p + 1) \\ \text{to each side.}\end{array}$$

$$0 = \boxed{} \qquad \begin{array}{l}\text{Divide each} \\ \text{side by 3.}\end{array}$$

$$0 = \boxed{} \qquad \text{Factor.}$$

$$\boxed{} = 0 \text{ or } \boxed{} = 0 \qquad \begin{array}{l}\text{Zero Product} \\ \text{Property}\end{array}$$

$$p = \boxed{} \qquad p = \boxed{} \qquad \begin{array}{l}\text{Solve each} \\ \text{equation.}\end{array}$$

Since $p = -1$ results in a zero in the denominator, eliminate -1.

Your Turn Solve $\dfrac{1}{x - 2} = \dfrac{2x + 1}{x^2 + 2x - 8} + \dfrac{2}{x + 4}.$

EXAMPLE Solve a Rational Inequality

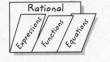
3 Solve $\dfrac{1}{3s} + \dfrac{2}{9s} < \dfrac{2}{3}$.

STEP 1 Values that make the denominator equal to 0 are
excluded from the denominator. For this inequality
the excluded value is 0.

STEP 2 Solve the related equation.

$$\dfrac{1}{3s} + \dfrac{2}{9s} = \dfrac{2}{3} \qquad \text{Related equation.}$$

$$\boxed{}\left(\dfrac{1}{3s} + \dfrac{2}{9s}\right) = \dfrac{2}{3}\boxed{} \qquad \text{Multiply each side}$$

by $\boxed{}$.

$$\boxed{} = \boxed{} \qquad \text{Simplify.}$$

$$\boxed{} = \boxed{} \qquad \text{Add.}$$

$$\boxed{} = \boxed{} \qquad \text{Divide each}$$

side by $\boxed{}$.

STEP 3 Draw vertical lines at the excluded value and at the
solution to separate the number line into regions.

Now test a sample value in each region to determine
if the values in the region satisfy the inequality.

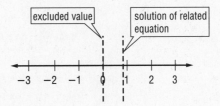

Test $s = -1$.

$$\dfrac{1}{3(-1)} + \dfrac{2}{9(-1)} \overset{?}{<} \dfrac{2}{3}$$

So, $s < 0$ is a solution.

Test $s = \dfrac{1}{3}$.

$$\dfrac{1}{3\left(\dfrac{1}{3}\right)} + \dfrac{2}{9\left(\dfrac{1}{3}\right)} \overset{?}{<} \dfrac{2}{3}$$

$$\boxed{} + \boxed{} \overset{?}{<} \dfrac{2}{3}$$

$$\boxed{} \not< \dfrac{2}{3}$$

So, $0 < s < \dfrac{5}{6}$ is not a solution.

Test $s = 1$.

$$\dfrac{1}{3(1)} + \dfrac{2}{9(1)} \overset{?}{<} \dfrac{2}{3}$$

$$\boxed{} + \boxed{} \overset{?}{<} \dfrac{2}{3}$$

$$\boxed{} < \dfrac{2}{3}$$

So, $s > \dfrac{5}{6}$ is a solution.

Your Turn   Solve $\dfrac{1}{x} + \dfrac{3}{5x} < \dfrac{2}{5}$.

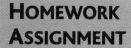

HOMEWORK ASSIGNMENT

Page(s):

Exercises:

BRINGING IT ALL TOGETHER

STUDY GUIDE

FOLDABLES	**VOCABULARY PUZZLEMAKER**	**BUILD YOUR VOCABULARY**
Use your **Chapter 9 Foldable** to help you study for your chapter test.	To make a crossword puzzle, word search, or jumble puzzle of the vocabulary words in Chapter 9, go to: www.glencoe.com/sec/math/ t_resources/free/index.php	You can use your completed **Vocabulary Builder** (page 202) to help you solve the puzzle.

9-1

Multiplying and Dividing Rational Expressions

1. Which expressions are complex fractions?

i. $\dfrac{7}{12}$ ii. $\dfrac{\frac{3}{8}}{\frac{5}{16}}$ iii. $\dfrac{r+5}{r-5}$ iv. $\dfrac{\frac{z+1}{z}}{z}$ v. $\dfrac{\frac{r^2-25}{9}}{\frac{r+5}{3}}$

Simplify each expression.

2. $\dfrac{6r^2s^3}{t^3} \cdot \dfrac{3s^2t^2}{12rst}$

3. $\dfrac{3y^2+3y-6}{4y-8} \div \dfrac{y^2-4}{2y^2-6y+4}$

9-2

Adding and Subtracting Rational Expressions

Find the LCM of each set of polynomials.

4. $4y, 9xy, 6y^2$

5. x^2-5x+6, x^3-4x^2+4x

Simplify each expression.

6. $\dfrac{6}{4ab} - \dfrac{3}{ab^2}$

7. $\dfrac{5q}{2p} + 8$

9-3

Graphing Rational Functions

For Exercises 8 and 9, refer to the two rational functions shown.

I.
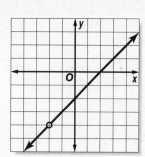

II.
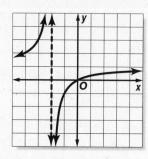

8. Graph I has a _____ at $x =$ ____.

9. Graph II has a _____ at $x =$ ____.

10. Graph $f(x) = \dfrac{3}{x - 2}$.

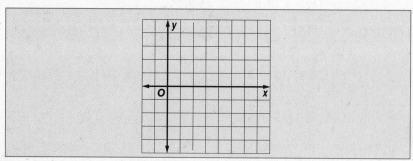

9-4

Direct, Joint, and Inverse Variation

11. Suppose y varies inversely as x. Find y when $x = 4$, if $y = 8$ when $x = 3$.

Which type of variation, direct or inverse, is represented by each graph?

12.

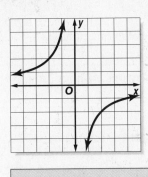

13.

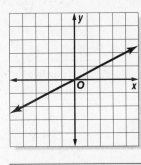

Classes of Functions

Match each graph below with the type of function it represents. Some types may be used more than once and others not at all.

a. square root	**b.** absolute value
c. rational	**d.** greatest integer
e. constant	**f.** identity

14.

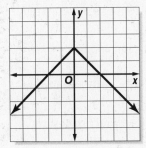

15.

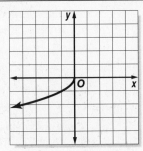

16.

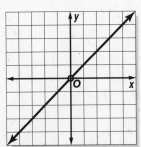

17.

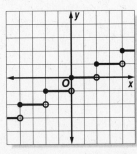

Solving Rational Equations and Inequalities

Solve each equation or inequality.

18. $\dfrac{1}{2x} + \dfrac{x+1}{x} = 4$

19. $\dfrac{y+1}{y-1} - \dfrac{2}{y-1} = \dfrac{y}{3}$

20. $\dfrac{3}{z+2} - \dfrac{6}{z} > 0$

ARE YOU READY FOR THE CHAPTER TEST?

Visit **algebra2.com** to access your textbook, more examples, self-check quizzes, and practice tests to help you study the concepts in Chapter 9.

Check the one that applies. Suggestions to help you study are given with each item.

☐ **I completed the review of all or most lessons without using my notes or asking for help.**

- You are probably ready for the Chapter Test.

- You may want take the Chapter 9 Practice Test on page 517 of your textbook as a final check.

☐ **I used my Foldable or Study Notebook to complete the review of all or most lessons.**

- You should complete the Chapter 9 Study Guide and Review on pages 513–516 of your textbook.

- If you are unsure of any concepts or skills, refer back to the specific lesson(s).

- You may also want to take the Chapter 9 Practice Test on page 517.

☐ **I asked for help from someone else to complete the review of all or most lessons.**

- You should review the examples and concepts in your Study Notebook and Chapter 9 Foldable.

- Then complete the Chapter 9 Study Guide and Review on pages 513–516 of your textbook.

- If you are unsure of any concepts or skills, refer back to the specific lesson(s).

- You may also want to take the Chapter 9 Practice Test on page 517.

Student Signature	Parent/Guardian Signature

Teacher Signature

Exponential and Logarithmic Relations

 Use the instructions below to make a Foldable to help you organize your notes as you study the chapter. You will see Foldable reminders in the margin of this Interactive Study Notebook to help you in taking notes.

Begin with four sheets of grid paper.

STEP 1 **Fold and Cut**
Fold in half along the width. On the first two sheets, cut along the fold at the ends. On the second two sheets, cut in the center of the fold as shown.

First Sheets

Second Sheets

STEP 2 **Fold and Label**
Insert first sheets through second sheets and align folds. Label pages with lesson numbers.

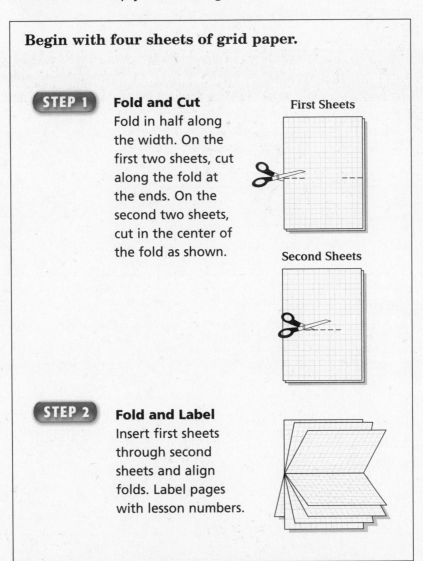

 NOTE-TAKING TIP: When you take notes, include personal experiences that relate to the lesson and ways in which what you have learned will be used in your daily life.

CHAPTER 10

This is an alphabetical list of new vocabulary terms you will learn in Chapter 10. As you complete the study notes for the chapter, you will see Build Your Vocabulary reminders to complete each term's definition or description on these pages. Remember to add the textbook page number in the second column for reference when you study.

Vocabulary Term	Found on Page	Definition	Description or Example
Change of Base Formula			
common logarithm [LAW-guh-RIH-thuhm]			
exponential decay [EHK-spuh-NEHN-chuhl]			
exponential equation			
exponential function			
exponential growth			
exponential inequality			

Vocabulary Term	Found on Page	Definition	Description or Example
logarithm			
logarithmic function [LAW-guh-RIHTH-mihk]			
natural base, *e*			
natural base exponential function			
natural logarithm			
natural logarithmic function			
rate of decay			
rate of growth			

Exponential Functions

Graph an Exponential Function

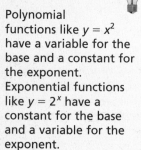

WHAT YOU'LL LEARN

- Graph exponential functions.

- Solve exponential equations and inequalities.

REMEMBER IT

Polynomial functions like $y = x^2$ have a variable for the base and a constant for the exponent. Exponential functions like $y = 2^x$ have a constant for the base and a variable for the exponent.

1 Sketch the graph of $y = 4^x$. Then state the function's domain and range.

Make a table of values. Connect the points to sketch a smooth curve.

x	y = 4ˣ
−2	
−1	
0	
1	
2	

The domain is _____, while the range is

_____.

Your Turn Sketch the graph of $y = 3^x$. Then state the function's domain and range.

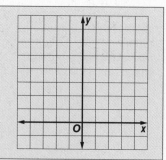

BUILD YOUR VOCABULARY (page 224)

In general, an equation of the form _____, where $a \neq 0$, $b > 0$, and $b \neq 1$, is called an **exponential function** with base b.

EXAMPLE Identify Exponential Growth and Decay

KEY CONCEPT

Exponential Growth and Decay

- If $a > 0$ and $b > 1$, the function $y = ab^x$ represents exponential growth.

- If $a > 0$ and $0 < b < 1$, the function $y = ab^x$ represents exponential decay.

2 Determine whether each function represents exponential *growth* or *decay*.

a. $y = (0.7)^x$ The function represents exponential

[_____], since the base, [_____],

is [_____].

b. $y = \frac{1}{2}(3)^x$ The function represents exponential

[_____], since the base, [____],

is [_____].

c. $y = 10\left(\frac{4}{3}\right)^x$ The function represents exponential

[_____], since the base, [____],

is [_____].

FOLDABLES

ORGANIZE IT

Use the page for Lesson 10-1. Write your own example of an exponential growth and an exponential decay function. Then sketch a graph of each function.

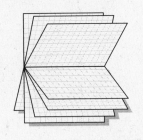

Your Turn   Determine whether each function represents exponential *growth* or *decay*.

a. $y = (0.5)^x$ [_____]

b. $y = \frac{1}{3}(2)^x$ [_____]

c. $y = 10\left(\frac{2}{5}\right)^x$ [_____]

BUILD YOUR VOCABULARY (page 224)

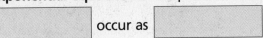

Exponential equations are equations in which

[_____] occur as [_____].

Exponential inequalities are inequalities involving

[_____] functions.

EXAMPLE Solve Exponential Equations

3 Solve $4^{9n-2} = 256$.

$4^{9n-2} = 256$ Original equation

$4^{9n-2} = $ Rewrite 256 as 4^4 so each side has the same base.

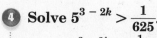 $= $ Property of Equality for Exponential Functions

Add 2 to each side.

$n = $ Divide each side by 9.

EXAMPLE Solve Exponential Inequalities

4 Solve $5^{3-2k} > \dfrac{1}{625}$.

$5^{3-2k} > \dfrac{1}{625}$ Original inequality

$5^{3-2k} > $ Rewrite $\dfrac{1}{625}$ as $\dfrac{1}{5^4}$ or 5^{-4}.

 $> $ Property of Inequality for Exponential Functions

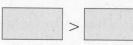 $> $ Subtract 3 from each side.

$k < $ Divide each side by -2.

Your Turn Solve each equation.

a. $2^{3x+1} = 32$

b. $5^{2x} = 25^{2x-1}$

c. $3^{3-2k} > \dfrac{1}{27}$

Logarithms and Logarithmic Functions

WHAT YOU'LL LEARN

- Evaluate logarithmic expressions.
- Solve logarithmic equations and inequalities.

KEY CONCEPT

Logarithm with Base b
Let b and x be positive numbers, $b \neq 1$. The *logarithm of x with base b* is denoted $\log_b x$ and is defined as the exponent y that makes the equation $b^y = x$ true.

EXAMPLE Logarithmic to Exponential Form

① **Write $\log_3 9 = 2$ in exponential form.**

$$\log_3 9 = 2 \longrightarrow \boxed{}$$

EXAMPLE Exponential to Logarithmic Form

② **Write $27^{\frac{1}{3}} = 3$ in logarithmic form.**

$$27^{\frac{1}{3}} = 3 \longrightarrow \boxed{}$$

Your Turn

Write each equation in exponential form.

a. $\log_2 8 = 3$

b. $\log_3 \dfrac{1}{9} = -2$

Write each equation in logarithmic form.

c. $3^4 = 81$

d. $81^{\frac{1}{2}} = 9$

EXAMPLE Evaluate Logarithmic Expressions

③ **Evaluate $\log_3 243$.**

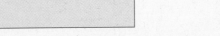

$\log_3 243 = y$		Let the logarithm equal y.
$\boxed{} = \boxed{}$		Definition of logarithm
$\boxed{} = \boxed{}$		$243 = 3^5$
$\boxed{} = y$		Property of Equality for Exponential Functions

Your Turn Evaluate $\log_{10} 1000$. $\boxed{}$

KEY CONCEPTS

Logarithmic to Exponential Inequality
If $b > 1$, $x > 0$, and $\log_b x > y$, then $x > b^y$.

If $b > 1$, $x > 0$, and $\log_b x < y$, then $0 < x < b^y$.

Property of Equality for Logarithmic Functions
If b is a positive number other than 1, then $\log_b x = \log_b y$ if and only if $x = y$.

Property of Inequality for Logarithmic Functions If $b > 1$, then $\log_b x > \log_b y$ if and only if $x > y$, and $\log_b x < \log_b y$ if and only if $x < y$.

EXAMPLE Solve a Logarithmic Equation

4 Solve $\log_8 n = \frac{4}{3}$.

$\log_8 n = \frac{4}{3}$ Original equation

$n = \boxed{}$ Definition of logarithm

$n = \boxed{}$ $8 = 2^3$

$n = \boxed{}$ Power of a Power

$n = \boxed{}$ Simplify.

Your Turn Solve $\log_{27} n = \frac{2}{3}$.

EXAMPLE Solve a Logarithmic Inequality

5 Solve $\log_6 x > 3$.

$\log_6 x > 3$ Original inequality

     Logarithmic to exponential inequality

     Simplify.

Your Turn Solve $\log_3 x < 2$. Check your solution.

HOMEWORK ASSIGNMENT

Page(s):

Exercises:

Properties of Logarithms

© Glencoe/McGraw-Hill

WHAT YOU'LL LEARN

- Simplify and evaluate expressions using the properties of logarithms.
- Solve logarithmic equations using the properties of logarithms.

KEY CONCEPTS

Product Property of Logarithms The logarithm of a product is the sum of the logarithms of its factors.

Quotient Property of Logarithms The logarithm of a quotient is the difference of the logarithms of the numerator and the denominator.

EXAMPLE Use the Product Property

1 Use $\log_5 2 \approx 0.4307$ to approximate the value of $\log_5 250$.

$\log_5 250 = \log_5 (5^3 \cdot 2)$ — Replace 250 with $5^3 \cdot 2$.

$= \boxed{} + \boxed{}$ — Product Property

$= \boxed{} + \boxed{}$ — Inverse Property of Exponents and Logarithms

$\approx \boxed{}$ — Replace $\log_5 2$.

EXAMPLE Use the Quotient Property

2 Use $\log_6 8 \approx 1.1606$ and $\log_6 32 \approx 1.9343$ to approximate the value of $\log_6 4$.

$\log_6 4 = \log_6\left(\dfrac{32}{8}\right)$ — Replace 4 with $\dfrac{32}{8}$.

$= \boxed{} - \boxed{}$ — Quotient Property

$\approx \boxed{} - \boxed{}$ — $\log_6 8 \approx 1.1606$ and $\log_6 32 \approx 1.9343$

or $\boxed{}$

Your Turn

a. Use $\log_2 3 \approx 1.5850$ to approximate the value of $\log_2 96$.

b. Use $\log_5 4 \approx 0.8614$ and $\log_5 32 \approx 2.1534$ to approximate the value of $\log_5 8$.

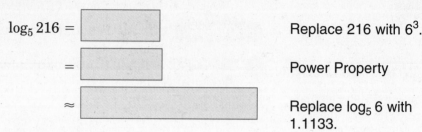

KEY CONCEPT

Power Property of Logarithms The logarithm of a power is the product of the logarithm and the exponent.

FOLDABLES Use the page for Lesson 10-3. Write your own examples that show the Product, Quotient, and Power Properties of Logarithms.

EXAMPLE Power Property of Logarithms

❸ **Given that $\log_5 6 \approx 1.1133$, approximate the value of $\log_5 216$.**

$\log_5 216 = $ [____]   Replace 216 with 6^3.

$= $ [____]   Power Property

$\approx $ [____]   Replace $\log_5 6$ with 1.1133.

Your Turn Given that $\log_4 6 \approx 1.2925$, approximate the value of $\log_4 1296$.

[____]

EXAMPLE Solve Equations Using Properties of Logarithms

❹ **Solve $4\log_2 x - \log_2 5 = \log_2 125$.**

$4\log_2 x - \log_2 5 = \log_2 125$ Original equation

[____] $- \log_2 5 = \log_2 125$ Power Property

[____] $= \log_2 125$ Quotient Property

[____] $= 125$ Property of Equality for Logarithmic Functions

[____] $= $ [____]   Multiply each side by 5.

$x = $ [____]   Take the 4th root of each side.

HOMEWORK ASSIGNMENT

Page(s): _____

Exercises: _____

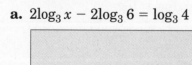 **Solve each equation.**

a. $2\log_3 x - 2\log_3 6 = \log_3 4$

[____]

b. $\log_2 x + \log_2 (x - 6) = 4$

[____]

Common Logarithms

EXAMPLE Find Common Logarithms

WHAT YOU'LL LEARN

- Solve exponential equations and inequalities using common logarithms.

- Evaluate logarithmic expressions using the Change of Base Formula.

1 Use a calculator to evaluate each expression to four decimal places.

a. log 6   **Keystrokes:** `LOG` 6 `ENTER` .7781512504

b. log 0.35 **Keystrokes:** `LOG` 0.35 `ENTER`

−.4559319556

Your Turn Use a calculator to evaluate each expression to four decimal places.

a. log 5

b. log 0.62

EXAMPLE Solve Exponential Equations Using Logarithms

REMEMBER IT

When solving an exponential equation using logarithms, the first step is often referred to as *taking the logarithm of each side.*

2 Solve $5^x = 62$.

$5^x = 62$ Original equation

$\boxed{} = \boxed{}$ Property of Equality for Logarithmic Functions

$\boxed{} = \boxed{}$ Power Property of Logarithms

$x = \boxed{}$ Divide each side by log 5.

$x \approx \boxed{}$ Use a calculator.

Your Turn Solve $3^x = 17$.

EXAMPLE Solve Exponential Inequalities Using Logarithms

3 Solve $2^{7x} > 3^{5x-3}$.

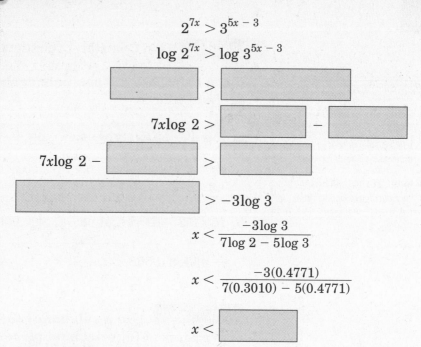

$$2^{7x} > 3^{5x-3}$$

$$\log 2^{7x} > \log 3^{5x-3}$$

$$\boxed{} > \boxed{}$$

$$7x\log 2 > \boxed{} - \boxed{}$$

$$7x\log 2 - \boxed{} > \boxed{}$$

$$\boxed{} > -3\log 3$$

$$x < \frac{-3\log 3}{7\log 2 - 5\log 3}$$

$$x < \frac{-3(0.4771)}{7(0.3010) - 5(0.4771)}$$

$$x < \boxed{}$$

Your Turn Solve $5^{3x} < 10^{x-2}$.

KEY CONCEPT

Change of Base Formula
For all positive numbers
a, b, and n, where $a \neq 1$
and $b \neq 1$,

$\log_a n = \dfrac{\log_b n}{\log_b a}$.

EXAMPLE Change of Base Formula

4 **Express $\log_3 18$ in terms of common logarithms. Then approximate its value to four decimal places.**

$$\log_3 18 = \boxed{} \approx \boxed{}$$

Your Turn Express $\log_5 16$ in terms of common logarithms. Then approximate its value to four decimal places.

HOMEWORK ASSIGNMENT

Page(s):

Exercises:

Base *e* and Natural Logarithms

WHAT YOU'LL LEARN

- Evaluate expressions involving the natural base and natural logarithms.

- Solve exponential equations and inequalities using natural logarithms.

BUILD YOUR VOCABULARY (page 225)

The [ ] number 2.71828 ... is referred to as the **natural base, e**.

An [ ] function with base [ ] is called a **natural base exponential function**.

The [ ] with base [ ] is called the **natural logarithm**.

The **natural logarithmic function**, $y = \ln x$, is the

[ ] of the natural base exponential function, $y = e^x$.

EXAMPLE Evaluate Natural Base Expressions

1 Use a calculator to evaluate each expression to four decimal places.

a. $e^{0.5}$

 Keystrokes: [2nd] $[e^x]$ 0.5 [ENTER]

 1.648721271 [ ]

b. e^{-8}

 Keystrokes: [2nd] $[e^x]$ −8 [ENTER]

 0.0003354626 [ ]

Your Turn Use a calculator to evaluate each expression to four decimal places.

a. $e^{0.3}$

b. e^{-2}

EXAMPLE Evaluate Natural Logarithmic Expressions

2 Use a calculator to evaluate each expression to four decimal places.

a. ln 3

Keystrokes: [LN] 3 [ENTER]

1.098612289 [   ]

b. ln $\frac{1}{4}$

Keystrokes: [LN] 1 ÷ 4 [ENTER]

−1.386294361 [   ]

Your Turn Use a calculator to evaluate each expression to four decimal places.

a. ln 2 [   ]   **b.** ln $\frac{1}{2}$ [   ]

EXAMPLE Write Equivalent Expressions

3 Write an equivalent exponential or logarithmic equation.

a. $e^x = 23$

$e^x = 23$ →

b. ln $x \approx 1.2528$

ln $x \approx 1.2528$ →

Your Turn Write an equivalent exponential or logarithmic equation.

a. $e^x = 6$ [   ]   **b.** ln $x = 2.25$ [   ]

EXAMPLE Inverse Property of Base e and Natural Logarithms

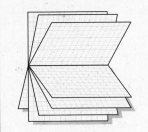

ORGANIZE IT

Use the page for Lesson 10-5. On the same grid, sketch the graph for $y = e^x$ and $y = \ln x$. Then write how you can tell that the two functions are inverses.

4 **Evaluate each expression.**

a. $e^{\ln 21}$

$e^{\ln 21} = $

b. $\ln e^{x^2 - 1}$

$\ln e^{x^2 - 1} = $

Your Turn **Evaluate each expression.**

a. $e^{\ln 7}$

b. $\ln e^{x + 3}$

EXAMPLE Solve Base e Equations

5 **Solve** $3e^{-2x} + 4 = 10$.

$3e^{-2x} + 4 = 10$		Original equation
☐ = ☐		Subtract 4 from each side.
☐ = ☐		Divide each side by 3.
$\ln e^{-2x} = \ln 2$		Property of Equality for Logarithms
☐ = ☐		Inverse Property of Exponents and Logarithms
$x = $ ☐		Divide each side by −2.
$x \approx $ ☐		Use a calculator.

HOMEWORK ASSIGNMENT

Page(s):

Exercises:

Your Turn Solve $2e^{-2x} + 5 = 15$.

Exponential Growth and Decay

WHAT YOU'LL LEARN

- Use logarithms to solve problems involving exponential decay.

- Use logarithms to solve problems involving exponential growth.

BUILD YOUR VOCABULARY (page 225)

The percent of decrease r is also referred to as the **rate of decay** in the equation for exponential decay of the form $y = a(1 - r)^t$.

EXAMPLE Exponential Decay of the Form $y = a(1 - r)t$

1 **CAFFEINE** A cup of coffee contains 130 milligrams of caffeine. If caffeine is eliminated from the body at a rate of 11% per hour, how long will it take for 90% of this caffeine to be eliminated from a person's body?

$$y = a(1 - r)^t$$ Exponential decay formula

$$13 = 130(1 - 0.11)^t$$ Replace y with 13, a with 130, and r with 0.11.

$$\boxed{} = \boxed{}$$ Divide each side by 130.

$$\log \boxed{} = \log \boxed{}$$ Property of Equality for Logarithms

$$\log 0.10 = \boxed{}$$ Power Property for Logarithms

$$\boxed{} = t$$ Divide each side by log 0.89.

$$\boxed{} \approx t$$ Use a calculator.

It will take approximately $\boxed{}$ hours for 90% of the caffeine to be eliminated from a person's body.

 Refer to Example 1. How long will it take for 80% of this caffeine to be eliminated from a person's body?

EXAMPLE Exponential Decay of the Form $y = ae^{-kt}$

2 GEOLOGY The half-life of Sodium-22 is 2.6 years.

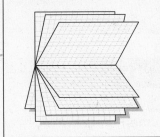
a. What is the value of k for Sodium-22?

$y = ae^{-kt}$	Exponential decay formula
$0.5a = ae^{-k(2.6)}$	Replace y and t.
$0.5 = e^{-2.6k}$	Divide each side by a.
$\ln \boxed{} = \ln \boxed{}$	Property of Equality for Logarithmic Functions
$\ln 0.5 = \boxed{}$	Inverse Property of Exponents and Logarithms
$\boxed{} = k$	Divide each side by -2.6.
$\boxed{} = k$	Use a calculator.

The constant k for Sodium-22 is $\boxed{}$. The equation for the decay of Sodium-22 is $y = ae^{-0.2666t}$, where t is years.

b. A geologist examining a meteorite estimates that it contains only about 10% as much Sodium-22 as it would have contained when it reached the surface of the Earth. How long ago did the meteorite reach the surface of the Earth?

$y = ae^{-0.2666t}$	Decay formula
$0.1a = ae^{-0.2666t}$	Replace y with $0.1a$.
$0.1 = e^{-0.2666t}$	Divide each side by a.
$\ln \boxed{} = \ln \boxed{}$	Property of Equality for Logarithms
$\ln \boxed{} = \boxed{}$	Inverse Property
$\boxed{} = t$	Divide each side by -0.2666.
$\boxed{} \approx t$	Use a calculator.

It was formed about $\boxed{}$ years ago.

Your Turn The half-life of radioactive iodine used in medical studies is 8 hours.

a. What is the value of k for radioactive iodine?

> (blank answer box)

b. A doctor wants to know when the amount of radioactive iodine in a patient's body is 20% of the original amount. When will this occur?

> (blank answer box)

BUILD YOUR VOCABULARY (page 225)

The percent of increase r is also referred to as the **rate of growth** in the equation for exponential growth of the form $y = a(1 + r)^t$.

EXAMPLE Exponential Growth of the Form $y = a(1 + r)^t$

3 The population of a city of one million is increasing at a rate of 3% per year. If the population continues to grow at this rate, in how many years will the population have doubled?

$y = a(1 + r)^t$	Growth formula
$2,000,000 = 1,000,000(1 + 0.03)^t$	Replace y, a, and r.
$\boxed{} = \boxed{}$	Divide each side by 1,000,000.
$\ln \boxed{} = \ln \boxed{}$	Property of Equality for Logarithms
$\ln 2 = \boxed{}$	Power Property of Logarithms
$\dfrac{\ln 2}{\ln 1.03} = t$	Divide.
$t \approx 23.45$	Use a calculator.

The population will have doubled in $\boxed{}$ years.

Your Turn The population of a city of 10,000 is increasing at a rate of 5% per year. If the population continues to grow at this rate, in how many years will the population have doubled?

> (blank answer box)

WRITE IT

How can you tell whether an exponential equation involves growth or decay?

_____

_____

_____

_____

HOMEWORK ASSIGNMENT

Page(s):

Exercises:

STUDY GUIDE

FOLDABLES™	VOCABULARY PUZZLEMAKER	**BUILD YOUR VOCABULARY**
Use your **Chapter 10 Foldable** to help you study for your chapter test.	To make a crossword puzzle, word search, or jumble puzzle of the vocabulary words in Chapter 10, go to: www.glencoe.com/sec/math/ t_resources/free/index.php	You can use your completed **Vocabulary Builder** (pages 224–225) to help you solve the puzzle.

10-1

Exponential Functions

Determine whether each function represents exponential *growth* or *decay*.

1. $y = 0.2(3)^x$

2. $y = 3\left(\dfrac{2}{5}\right)^x$

3. $y = 0.4(1.01)^x$

4. Simplify $4^x \cdot 4^{2x}$.

5. Solve $25^{y+3} = \left(\dfrac{1}{5}\right)^y$.

10-2

Logarithms and Logarithmic Functions

6. What is the inverse of the function $y = 5^x$?

7. What is the inverse of the function $y = \log_{10} x$?

8. Evaluate $\log_{27} 9$.

9. Solve $\log_8 x = -\dfrac{1}{3}$.

10-3

Properties of Logarithms

State whether each of the following equations is *true* or *false*. If the statement is true, name the property of logarithms that is illustrated.

10. $\log_3 10 = \log_3 30 - \log_3 3$

11. $\log_4 12 = \log_4 4 + \log_4 8$

12. $\log_2 81 = 2\log_2 9$

Solve each equation.

13. $\log_5 14 - \log_5 (2x) = \log_5 21$ **14.** $\log_2 x + \log_2 (x + 2) = 3$

10-4

Common Logarithms

Match each expression from the first column with an expression from the second column that has the same value.

15. $\log_2 2$

16. $\log 12$

17. $\log_3 1$

18. $\log_5 \dfrac{1}{5}$

19. $\log 1000$

a. $\log_4 1$

b. $\log_2 8$

c. $\log_{10} 12$

d. $\log_5 5$

e. $\log 0.1$

20. Solve $6^{2x-1} = 2^{5x}$. Round to four decimal places.

21. Express $\log_8 5$ in terms of common logarithms. Then approximate its value to four decimal places.

10-5

Base *e* and Natural Logarithms

Match each expression from the first column with its value in the second column. Some choices may be used more than once or not at all.

22. $e^{\ln 5}$

23. $\ln 1$

24. $e^{\ln e}$

25. $\ln e^5$

26. $\ln e$

27. $\ln \left(\dfrac{1}{e}\right)$

> **I.** 1
>
> **II.** 10
>
> **III.** -1
>
> **IV.** 5
>
> **V.** 0
>
> **IV.** *e*

28. Solve $\ln (x - 8) = 5$. Round to four decimal places.

29. Evaluate $e^{\ln 5.2}$.

10-6

Exponential Growth and Decay

State whether each equation represents exponential *growth* or *decay*.

30. $y = 5e^{0.15t}$

31. $y = 1000(1 - 0.05)^t$

32. $y = 0.3e^{-1200t}$

33. $y = 2(1 + 0.0001)^t$

34. LeRoy bought a lawn mower for $1,200. It is expected to depreciate at a rate of 20% per year. What will be the value of the lawn mower in 5 years?

35. The population of a school has increased at a steady rate each year from 375 students to 580 students in 8 years. Find the annual rate of growth.

ARE YOU READY FOR THE CHAPTER TEST?

Math Online

Visit **algebra2.com** to access your textbook, more examples, self-check quizzes, and practice tests to help you study the concepts in Chapter 10.

Check the one that applies. Suggestions to help you study are given with each item.

☐ **I completed the review of all or most lessons without using my notes or asking for help.**

- You are probably ready for the Chapter Test.

- You may want to take the Chapter 10 Practice Test on page 571 of your textbook as a final check.

☐ **I used my Foldable or Study Notebook to complete the review of all or most lessons.**

- You should complete the Chapter 10 Study Guide and Review on pages 566–570 of your textbook.

- If you are unsure of any concepts or skills, refer back to the specific lesson(s).

- You may also want to take the Chapter 10 Practice Test on page 571 of your textbook.

☐ **I asked for help from someone else to complete the review of all or most lessons.**

- You should review the examples and concepts in your Study Notebook and Chapter 10 Foldable.

- Then complete the Chapter 10 Study Guide and Review on pages 566–570 of your textbook.

- If you are unsure of any concepts or skills, refer back to the specific lesson(s).

- You may also want to take the Chapter 10 Practice Test on page 571 of your textbook.

Student Signature   Parent/Guardian Signature

Teacher Signature

Sequences and Series

 Use the instructions below to make a Foldable to help you organize your notes as you study the chapter. You will see Foldable reminders in the margin of this Interactive Study Notebook to help you in taking notes.

Begin with one sheet of 11" × 17" paper and four sheets of notebook paper.

STEP 1 **Fold and Cut**
Fold the short sides of the 11" × 17" paper to meet in the middle.

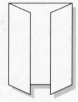

STEP 2 **Staple and Label**
Fold the notebook paper in half lengthwise. Insert two sheets of notebook paper in each tab and staple edges. Label with lesson numbers.

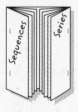

 NOTE-TAKING TIP: When you take notes, write questions you have about the lessons in the margin of your notes. Then include the answers to these questions as you work through the lesson.

BUILD YOUR VOCABULARY

This is an alphabetical list of new vocabulary terms you will learn in Chapter 11. As you complete the study notes for the chapter, you will see Build Your Vocabulary reminders to complete each term's definition or description on these pages. Remember to add the textbook page number in the second column for reference when you study.

Vocabulary Term	Found on Page	Definition	Description or Example
arithmetic mean [AR-ihth-MEH-tihk]			
arithmetic sequence			
arithmetic series			
Binomial Theorem			
common difference			
common ratio			
factorial			
Fibonacci sequence fih-buh-NAH-chee			
geometric mean			

Vocabulary Term	Found on Page	Definition	Description or Example
geometric sequence			
geometric series			
index of summation			
inductive hypothesis			
infinite geometric series			
iteration [IH-tuh-RAY-shuhn]			
mathematical induction			
partial sum			
Pascal's triangle [pas-KALZ]			
recursive formula [rih-KUHR-sihv]			
sigma notation [SIHG-muh]			

Arithmetic Sequences

WHAT YOU'LL LEARN

• Use arithmetic sequences.

• Find arithmetic means.

BUILD YOUR VOCABULARY (page 246)

An **arithmetic sequence** is a sequence in which each

[   ] after the first is found by [   ] a

constant, called the **common difference** d, to the

[   ] term.

KEY CONCEPT

nth Term of an Arithmetic Sequence The nth term a_n of an arithmetic sequence with first term a_1 and common difference d is given by $a_n = a_1 + (n - 1)d$, where n is any positive integer.

FOLDABLES On the tab for Lesson 11-1, write your own arithmetic sequence.

EXAMPLE Find the Next Terms

1 Find the next four terms of the arithmetic sequence $-8, -6, -4, \ldots$.

Find the common difference d by subtracting 2 consecutive terms.

$-6 - (-8) = $ [   ] and $-4 - (-6) = $ [   ] So, $d = $ [   ].

Now add 2 to the third term of the sequence and then continue adding 2 until the next four terms are found.

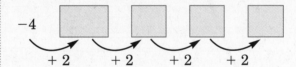

-4 [   ] [   ] [   ] [   ]
  $+2$ $+2$ $+2$ $+2$

EXAMPLE Write an Equation for the nth Term

2 Write an equation for the nth term of the arithmetic sequence $-8, -6, -4, \ldots$.

In this sequence, $a_1 = -8$ and $d = 2$. Use the nth formula to write an equation.

$a_n = a_1 + (n - 1)d$ Formula for the nth term

$a_n = $ [   ] $a_1 = -8, d = 2$

$a_n = $ [   ]   Distributive Property

$a_n = $ [   ]   Simplify.

REVIEW IT

When finding the value of an expression, you must follow the order of operations. Briefly list the order of operations. *(Lesson 1-1)*

_____

_____

_____

_____

_____

Your Turn

a. Find the next four terms of the arithmetic sequence 5, 3, 1, ...

b. Write an equation for the nth term of the arithmetic sequence 5, 3, 1, ...

BUILD YOUR VOCABULARY (page 246)

The terms between any two nonsuccessive terms of an

_____ sequence are called **arithmetic means**.

EXAMPLE Find Arithmetic Means

③ **Find the three arithmetic means between 21 and 45.**

You can use the nth term formula to find the common difference. In the sequence 21, __, __, __, 45, . . . , $a_1 = 21$ and $a_5 = 45$.

$$a_n = a_1 + (n-1)d \qquad \text{Formula for the } n\text{th term}$$

$$a_5 = \boxed{} \qquad n = 5,\ a_1 = 21$$

$$\boxed{} = \boxed{} \qquad a_5 = 45$$

$$\boxed{} = \boxed{} \qquad \text{Subtract 21 from each side.}$$

$$\boxed{} = d \qquad \text{Divide each side by 4.}$$

Now use the value of d to find the three arithmetic means.

21 $\boxed{}$ $\boxed{}$ $\boxed{}$

+ 6   + 6   + 6

Your Turn Find the three arithmetic means between 13 and 25.

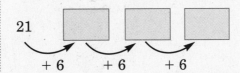

HOMEWORK ASSIGNMENT

Page(s): _____

Exercises: _____

Arithmetic Series

BUILD YOUR VOCABULARY (page 246)

WHAT YOU'LL LEARN

- Find sums of arithmetic series.
- Use sigma notation.

A series is an indicated **sum** of the [ ] of a sequence. Since 18, 22, 26, 30 is an **arithmetic** sequence, $18 + 22 + 26 + 30$ is an **arithmetic series**.

EXAMPLE Find the Sum of an Arithmetic Series

1 Find the sum of the first 20 even numbers, beginning with 2.

The series is $2 + 4 + 6 + \ldots + 40$. Since $a_1 = 2$, $a_{20} = 40$, and $d = 2$, you can use either sum formula for this series.

KEY CONCEPT

Sum of an Arithmetic Series The sum S_n of the first n terms of an arithmetic series is given by

$S_n = \frac{n}{2}[2a_1 + (n-1)d]$ or

$S_n = \frac{n}{2}(a_1 + a_n)$.

Method 1

$S_n = \frac{n}{2}(a_1 + a_n)$ Sum formula

$S_{20} = $ [ ] $n = 20$, $a_1 = 2$, $a_{20} = 40$

$S_{20} = $ [ ]   Simplify.

$S_{20} = $ [ ]   Multiply.

Method 2

$S_n = \frac{n}{2}[2a_1 + (n-1)d]$ Sum formula

$S_{20} = $ [ ] $n = 20$, $a_1 = 2$, $d = 2$

$S_{20} = $ [ ]   Simplify.

$S_{20} = $ [ ]   Multiply.

Your Turn Find the sum of the first 15 counting numbers, beginning with 1.

[ ]

EXAMPLE Find the First Three Terms

❷ Find the first three terms of an arithmetic series in which $a_1 = 14$, $a_n = 29$, and $S_n = 129$.

STEP 1 Since you know a_1, a_n, and S_n, use $S_n = \frac{n}{2}(a_1 + a_n)$ to find n.

$$S_n = \frac{n}{2}(a_1 + a_n)$$

$$\boxed{} = \frac{n}{2}\boxed{}$$

$$\boxed{} = \boxed{}$$

$$\boxed{} = n$$

STEP 2 Find d.

$$a_n = a_1 + (n - 1)d$$

$$\boxed{} = \boxed{} + \boxed{}\,d$$

$$\boxed{} = \boxed{}$$

$$\boxed{} = d$$

STEP 3 Use d to determine a_2 and a_3.

$$a_2 = \boxed{} + \boxed{} \quad \text{or} \quad \boxed{}$$

$$a_3 = \boxed{} + \boxed{} \quad \text{or} \quad \boxed{}$$

Your Turn Find the first three terms of an arithmetic series in which $a_1 = 11$, $a_n = 31$ and $S_n = 105$.

FOLDABLES

ORGANIZE IT

Use the tab for Lesson 11-2. Write an example of an arithmetic sequence and an arithmetic series. Then explain the difference between the two.

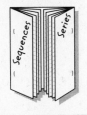

© Glencoe/McGraw-Hill

BUILD YOUR VOCABULARY (page 247)

A concise notation for writing out a series is called **sigma notation**. The series $3 + 6 + 9 + 12 + \ldots + 30$ can be expressed as $\sum_{n=1}^{10} 3n$. The variable, in this case n, is called the **index of summation**.

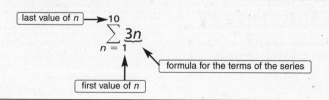

last value of n → 10

$$\sum_{n=1} 3n$$

$n = 1$

first value of n

formula for the terms of the series

EXAMPLE Evaluate a Sum in Sigma Notation

3 Evaluate $\displaystyle\sum_{k=3}^{10} (2k + 1)$.

Method 1 Find the terms by replacing k with $3, 4, \ldots, 10$. Then add.

$$\sum_{k=3}^{10} (2k + 1)$$

$$= [2(3) + 1] + [2(4) + 1] + [2(5) + 1] + [2(6) + 1] + [2(7) + 1] + [2(8) + 1] + [2(9) + 1] + [2(10) + 1]$$

$= \boxed{} + \boxed{} + \boxed{} + \boxed{} + \boxed{} +$

$\boxed{} + \boxed{} + \boxed{}$

$= \boxed{}$

Method 2 Since the sum is an arithmetic series, use the formula $S_n = \dfrac{n}{2}(a_1 + a_n)$. There are 8 terms, $a_1 = 2(3) + 1$ or 7, and $a_8 = 2(10) + 1$ or 21.

$$S_n = \boxed{} = 4(28) \text{ or } \boxed{}$$

© Glencoe/McGraw-Hill

HOMEWORK ASSIGNMENT

Page(s):

Exercises:

Your Turn Evaluate $\displaystyle\sum_{i=5}^{10} (2i + 3)$.

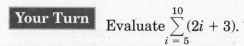

Geometric Sequences

WHAT YOU'LL LEARN

- Use geometric sequences.
- Find geometric means.

BUILD YOUR VOCABULARY (pages 246–247)

A **geometric sequence** is a sequence in which each term after the first is found by multiplying the previous term by a constant r called the **common ratio**.

The missing term(s) between two nonsuccessive terms of a geometric sequence are called **geometric means**.

EXAMPLE Find the Next Term

1 Find the missing term in the geometric sequence 324, 108, 36, 12, _____.

Since $\frac{108}{324} = \frac{1}{3}$, $\frac{36}{108} = \frac{1}{3}$, and $\frac{12}{36} = \frac{1}{3}$, the sequence has the

common ratio of $\boxed{}$.

Find the missing term.　　　$12\left(\frac{1}{3}\right) = \boxed{}$

Your Turn Find the missing term in the geometric sequence 100, 50, 25, _____.

EXAMPLE Find a Particular Term

KEY CONCEPT

nth Term of a Geometric Sequence The nth term a_n of a geometric sequence with first term a_1 and common ratio r is given by $a_n = a_1 \cdot r^{n-1}$, where n is any positive integer.

2 Find the sixth term of a geometric sequence for which $a_1 = -3$ and $r = -2$.

$a_n = a_1 \cdot r^{n-1}$ 　　　　Formula for the nth term

$a_6 = \boxed{} \cdot \boxed{}$ 　　$n = 6$, $a_1 = -3$, $r = -2$

$a_6 = \boxed{} \cdot \boxed{}$ 　　$(-2)^5 = -32$

$a_6 = \boxed{}$ 　　　　Multiply.

Your Turn Find the fifth term of a geometric sequence for which $a_1 = 6$ and $r = 2$.

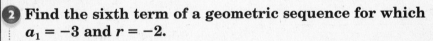

EXAMPLE Find a Term Given the Fourth Term and the Ratio

3 Find the seventh term of a geometric sequence for which $a_3 = 96$ and $r = 2$.

First find the value of a_1.

$$a_n = a_1 r^{n-1}$$

$$\boxed{} = \boxed{}$$

$$96 = a_1(2)^2$$

$$24 = a_1$$

Now find a_7.

$$a_n = a_1 r^{n-1}$$

$$a_7 = \boxed{}$$

$$a_7 = \boxed{}$$

Your Turn Find the sixth term of a geometric sequence for which $a_4 = 27$ and $r = 3$.

$$\boxed{}$$

FOLDABLES

ORGANIZE IT

Use the tab for Lesson 11-3. Write your own example of an arithmetic sequence and a geometric sequence. Then explain the difference between the two.

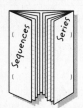

EXAMPLE Find Geometric Means

4 Find three geometric means between 3.12 and 49.92.

In the sequence a_1 is 3.12 and a_5 is 49.92.

$$a_n = a_1 r^{n-1}$$ Formula for the nth term

$$a_5 = \boxed{}$$ $n = 5$, $a_1 = 3.12$

$$\boxed{} = \boxed{}$$ $a_5 = 49.92$

$$16 = r^4$$ Divide by 3.12.

$$\pm 2 = r$$ Take the fourth root of each side.

Use each value of r to find three geometric means.

$r = 2$

$$a_2 = 3.12(2) \text{ or } \boxed{}$$

$$a_3 = 6.24(2) \text{ or } \boxed{}$$

$$a_4 = 12.48(2) \text{ or } \boxed{}$$

$r = -2$

$$a_2 = 3.12(-2) \text{ or } \boxed{}$$

$$a_3 = -6.24(-2) \text{ or } \boxed{}$$

$$a_4 = 12.48(-2) \text{ or } \boxed{}$$

HOMEWORK ASSIGNMENT

Page(s):

Exercises:

Your Turn Find three geometric means between 12 and 0.75.

$$\boxed{}$$

Geometric Series

EXAMPLE Find the Sum of the First *n* Terms

WHAT YOU'LL LEARN

- Find sums of geometric series.

- Find specific terms of geometric series.

KEY CONCEPT

Sum of a Geometric Series The sum S_n of the first *n* terms of a geometric series is given by $S_n = \dfrac{a_1 - a_1 r^n}{1 - r}$ or $S_n = \dfrac{a_1(1 - r^n)}{1 - r}$, where $r \neq 1$.

REMEMBER IT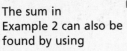

The sum in Example 2 can also be found by using $S_n = \dfrac{a_1 - a_1 r^n}{1 - r}$.

1 GENEALOGY **How many direct ancestors would a person have after 8 generations?**

Counting two parents, four grandparents, eight great grandparents, and so on gives you a geometric series with $a_1 = 2$, $r = 2$, and $n = 8$.

$S_n = \dfrac{a_1(1 - r^n)}{1 - r}$ Sum formula

$= $ [ ] $n = 8$, $a_1 = 2$, $r = 2$

$= $ [ ]   Use a calculator.

EXAMPLE Evaluate a Sum Written in Sigma Notation

2 Evaluate $\displaystyle\sum_{n=1}^{12} 3 \cdot 2^{n-1}$.

The sum is a geometric series.

$S_n = \dfrac{a_1(1 - r^n)}{1 - r}$ Sum formula

$S_{12} = $ [ ] $n = 12$, $a_1 = 3$, $r = 2$

$S_{12} = \dfrac{3(4095)}{1}$ $2^{12} = 4096$

$S_{12} = $ [ ]   Simplify.

Your Turn

a. How many direct ancestors would a person have after 7 generations?

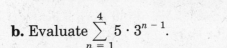

b. Evaluate $\displaystyle\sum_{n=1}^{4} 5 \cdot 3^{n-1}$.

EXAMPLE Use the Alternate Formula for a Sum

③ Find the sum of a geometric series for which $a_1 = 7776$, $a_n = 6$, and $r = -\frac{1}{6}$.

Since you do not know the value of n use $S_n = \frac{a_1 - a_n r}{1 - r}$.

$S_n = \frac{a_1 - a_n r}{1 - r}$ Alternate sum formula

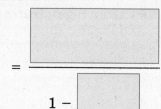

$= \dfrac{\boxed{}}{1 - \boxed{}}$ $a_1 = 7776$, $a_n = 6$, and $r = -\frac{1}{6}$

$= \dfrac{7777}{\frac{7}{6}}$ or $\boxed{}$ Simplify.

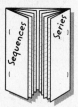
Your Turn Find the sum of a geometric series for which $a_1 = 64$, $a_n = 729$, and $r = -1.5$.

EXAMPLE Find the First Term of a Series

④ Find a_1 in a geometric series for which $S_8 = 765$ and $r = 2$.

$S_n = \dfrac{a_1(1 - r^n)}{1 - r}$ Sum formula

$\boxed{} = \boxed{}$ $S_8 = 765$, $r = 2$, and $n = 8$

$765 = 255a_1$ Simplify.

$\boxed{} = a_1$ Divide each side by 255.

HOMEWORK ASSIGNMENT

Page(s): _____

Exercises: _____

_____

_____

Your Turn Find a_1 in a geometric series for which $S_6 = 364$ and $r = 3$.

11–5 Infinite Geometric Series

BUILD YOUR VOCABULARY (page 247)

WHAT YOU'LL LEARN

- Find the sum of an infinite geometric series.
- Write repeating decimals as fractions.

If a geometric series has no last term, it is called an **infinite geometric series**. For infinite series, S_n is called a **partial sum** of the series.

EXAMPLE Sum of an Infinite Geometric Series

KEY CONCEPT

Sum of an Infinite Geometric Series The sum S of an infinite geometric series with $-1 < r < 1$ given by $S = \frac{a_1}{1-r}$.

1. Find the sum of $3 - \frac{3}{2} + \frac{3}{4} - \frac{3}{8} + \ldots$, if it exists.

$a_1 = \boxed{}$ and $a_2 = \boxed{}$, so $r = \dfrac{-\frac{3}{2}}{3}$ or $\boxed{}$.

Since $\boxed{} < 1$, the sum exists.

Now use the formula for the sum of an infinite geometric series.

$S = \dfrac{a_1}{1-r}$ Sum formula

$= \dfrac{\boxed{}}{1 - \boxed{}}$ $a_1 = 3, \ r = -\frac{1}{2}$

$= \dfrac{\boxed{}}{\boxed{}}$ or $\boxed{}$ Simplify.

Your Turn Find the sum of each infinite geometric series, if it exists.

a. $2 + 4 + 8 + 16 + \ldots$

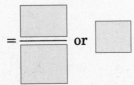

b. $1 + \frac{1}{2} + \frac{1}{4} + \ldots$

EXAMPLE Infinite Series in Sigma Notation

2 Evaluate $\displaystyle\sum_{n=1}^{\infty} 5\left(\frac{1}{2}\right)^{n-1}$.

In this infinite geometric series, $a_1 = 5$ and $r = \frac{1}{2}$.

$S = \dfrac{a_1}{1-r}$ Sum formula

$ = \dfrac{5}{1 - \frac{1}{2}}$ $a_1 = 5$, $r = \frac{1}{2}$

$ = $ [ ] or [ ] Simplify.

Your Turn Evaluate $\displaystyle\sum_{n=1}^{\infty} 2\left(\frac{1}{3}\right)^{n-1}$.

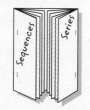
EXAMPLE Write a Repeating Decimal as a Fraction

3 Write $0.\overline{25}$ as a fraction.

$S = 0.\overline{25}$ Label the given decimal.

$S = 0.25252525\ldots$ Repeating decimal

$100S = $ [ ] Multiply each side by 100.

$99S = $ [ ] Subtract the second equation from the third.

$S = $ [ ] Divide each side by 99.

Your Turn Write $0.\overline{37}$ as a fraction.

© Glencoe/McGraw-Hill

HOMEWORK ASSIGNMENT

Page(s):

Exercises:

Recursion and Special Sequences

WHAT YOU'LL LEARN

- Recognize and use special sequences.
- Iterate functions.

FOLDABLES

ORGANIZE IT
Use the tab for Lesson 11-6. Describe the pattern in the sequence 1, 2, 6, 24, 120, ... Then find the next three terms of the sequence.

Sequences   Series

BUILD YOUR VOCABULARY (pages 246–247)

The sequence 1, 1, 2, 3, 5, 8, 13, ... , where each term in the sequence after the second is the sum of the two previous terms, is called the **Fibonacci sequence**.

The formula $a_n = a_{n-2} + a_{n-1}$ is an example of a **recursive formula**.

EXAMPLE Use a Recursive Formula

1 Find the first five terms of the sequence in which $a_1 = 5$ and $a_{n+1} = 2a_n + 7$, $n \geq 1$.

$a_{n+1} = 2a_n + 7$	Recursive formula
$a_{1+1} = 2a_1 + 7$	$n = 1$
$a_2 = 2\left(\boxed{}\right) + 7$ or $\boxed{}$	$a_1 = 5$
$a_{2+1} = 2a_2 + 7$	$n = 2$
$a_3 = 2\left(\boxed{}\right) + 7$ or $\boxed{}$	$a_2 = 17$
$a_{3+1} = 2a_3 + 7$	$n = 3$
$a_4 = 2\left(\boxed{}\right) + 7$ or $\boxed{}$	$a_3 = 41$
$a_{4+1} = 2a_4 + 7$	$n = 4$
$a_5 = 2\left(\boxed{}\right) + 7$ or $\boxed{}$	$a_4 = 89$

Your Turn Find the first five terms of the sequence in which $a_1 = 2$ and $a_{n+1} = 3a_n + 2$, $n \geq 1$.

BUILD YOUR VOCABULARY (page 247)

Iteration is the process of composing a [   ] with itself repeatedly.

EXAMPLE Iterate a Function

2 Find the first three iterates x_1, x_2, and x_3 of the function $f(x) = 3x - 1$ for an initial value of $x_0 = 5$.

To find the first iterate x_1, find the value of the function for $x_0 = 5$.

$x_1 = f(x_0)$ Iterate the function.

$= f$ [   ] $x_0 = 5$

$=$ [   ] or [   ]   Simplify.

To find the second iterate x_2, substitute x_1 for x.

$x_2 = f(x_1)$ Iterate the function.

$= f$ [   ] $x_1 = 14$

$=$ [   ] or [   ]   Simplify.

Substitute x_2 for x to find the third iterate.

$x_3 = f(x_2)$ Iterate the function.

$= f$ [   ] $x_2 = 41$

$=$ [   ] or [   ]   Simplify.

REVIEW IT

If $f(x) = 4x + 10$ and $g(x) = 2x^2 - 5$, find $f(g(x))$. (Lesson 7-7)

_____

_____

_____

_____

_____

Your Turn Find the first three iterates x_1, x_2, and x_3 of the function $f(x) = 2x + 1$ for an initial value of $x_0 = 2$.

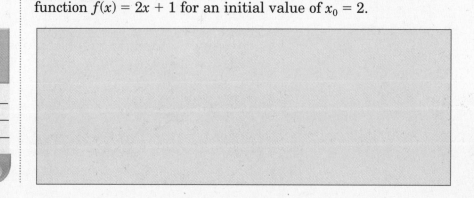

HOMEWORK ASSIGNMENT

Page(s): _____

Exercises: _____

_____

The Binomial Theorem

WHAT YOU'LL LEARN

- Use Pascal's triangle to expand powers of binomials.

- Use the Binomial Theorem to expand powers of binomials.

BUILD YOUR VOCABULARY (pages 246–247)

The ⬜ in powers of ⬜ form a pattern that is often displayed in a triangular formation known as **Pascal's triangle**.

The factors in the coefficients of binomial ⬜ involve special products called **factorials**.

EXAMPLE Use Pascal's Triangle

1 **Expand** $(p + q)^5$.

Write the row of Pascal's triangle corresponding to $n = 5$.

⬜ ⬜ ⬜ ⬜ ⬜ ⬜

Use the patterns of a binomial expansion and the coefficients to write the expansion of $(p + q)^5$.

$(p + q)^5$

$=$ ⬜

$=$ ⬜

FOLDABLES

ORGANIZE IT

Use the tab for Lesson 11-7. Refer to Example 1, and describe what happens to the exponents for p as the exponents for q increase.

EXAMPLE Use the Binomial Theorem

2 **Expand** $(t - s)^8$.

The expression will have nine terms. Use the sequence $1, \dfrac{8}{1}$, $\dfrac{8 \cdot 7}{1 \cdot 2}, \dfrac{8 \cdot 7 \cdot 6}{1 \cdot 2 \cdot 3}, \dfrac{8 \cdot 7 \cdot 6 \cdot 5}{1 \cdot 2 \cdot 3 \cdot 4}$ to find the coefficients for the first five terms. Use symmetry to find the remaining coefficients.

$(t - s)^8 = 1t^8(-s)^0 + \dfrac{8}{1}t^7(-s) + \dfrac{8 \cdot 7}{1 \cdot 2}t^6(-s)^2 + \dfrac{8 \cdot 7 \cdot 6}{1 \cdot 2 \cdot 3}t^5(-s)^3 +$

$\dfrac{8 \cdot 7 \cdot 6 \cdot 5}{1 \cdot 2 \cdot 3 \cdot 4}t^4(-s)^4 + \ldots + 1t^0(-s)^8$

$= t^8 - 8t^7s + 28t^6s^2 - 56t^5s^3 + 70t^4s^4 - 56t^3s^5 + 28t^2s^6$

$- 8ts^7 + s^8$

KEY CONCEPT

Binomial Theorem If n is a nonnegative integer, then

$(a + b)^n$

$= 1a^nb^0 + \dfrac{n}{1}a^{n-1}b^1 +$

$\dfrac{n(n-1)}{1 \cdot 2}a^{n-2}b^2 +$

$\dfrac{n(n-1)(n-2)}{1 \cdot 2 \cdot 3}a^{n-3}b^3 +$

$\ldots + 1a^0b^n.$

Your Turn

a. Expand $(x + y)^6$.

b. Expand $(x - y)^4$.

EXAMPLE Use a Factorial Form for the Binomial Theorem

③ Expand $(3x - y)^4$.

$$(3x - y)^4 =$$

KEY CONCEPT

Binomial Theorem, Factored Form

$(a + b)^n$

$= \dfrac{n!}{n!0!} a^n b^0 +$

$\dfrac{n!}{(n-1)!1!} a^{n-1} b^1 +$

$\dfrac{n!}{(n-2)!2!} a^{n-2} b^2 +$

$\ldots + \dfrac{n!}{0!n!} a^0 b^n +$

$= \displaystyle\sum_{k=0}^{n} \dfrac{n!}{(n-k)!k!} a^{n-k} b^k$

$$= \dfrac{4!}{4!0!}(3x)^4(-y)^0 + \dfrac{4!}{3!1!}(3x)^3(-y)^1 + \dfrac{4!}{2!2!}(3x)^2(-y)^2 +$$

$$\dfrac{4!}{1!3!}(3x)^1(-y)^3 + \dfrac{4!}{0!4!}(3x)^0(-y)^4$$

$$= \dfrac{4 \cdot 3 \cdot 2 \cdot 1}{4 \cdot 3 \cdot 2 \cdot 1 \cdot 1}(3x)^4 + \dfrac{4 \cdot 3 \cdot 2 \cdot 1}{3 \cdot 2 \cdot 1 \cdot 1}(3x)^3(-y) +$$

$$\dfrac{4 \cdot 3 \cdot 2 \cdot 1}{2 \cdot 1 \cdot 2 \cdot 1}(3x)^2 y^2 + \dfrac{4 \cdot 3 \cdot 2 \cdot 1}{1 \cdot 3 \cdot 2 \cdot 1}(3x)^1(-y^3) +$$

$$\dfrac{4 \cdot 3 \cdot 2 \cdot 1}{1 \cdot 4 \cdot 3 \cdot 2 \cdot 1}(3x)^0 y^4$$

$$=$$

Your Turn Expand $(2x + y)^4$.

HOMEWORK ASSIGNMENT

Page(s):

Exercises:

Proof and Mathematical Induction

WHAT YOU'LL LEARN

- Prove statements using mathematical induction.
- Disprove statements by finding a counterexample.

KEY CONCEPT

Mathematical Induction
STEP 1 Show that the statement is true for some integer n.

STEP 2 Assume that the statement is true for some positive integer k, where $k \geq n$. This assumption is called the **inductive hypothesis**.

STEP 3 Show that the statement is true for the next integer $k + 1$.

BUILD YOUR VOCABULARY (page 247)

Mathematical induction is used to ⬚ statements about ⬚ integers.

EXAMPLE Summation Formula

1 Prove that $1 + 3 + 5 + \ldots + (2n - 1) = n^2$.

STEP 1 When $n = 1$, the left side of the given equation is

⬚ – ⬚ or ⬚ . The right side is ⬚ or

⬚ . Thus, the equation is true for $n = 1$.

STEP 2 Assume $1 + 3 + 5 + \ldots + (2k - 1) = k^2$ for a positive integer k.

STEP 3 Show that the given equation is true for

$n = $ ⬚ .

$$1 + 3 + 5 + \ldots + (2k - 1) + (2(k + 1) - 1)$$

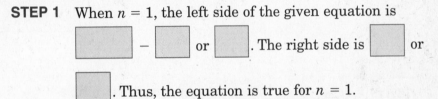

= ⬚   Add $(2(k + 1) - 1)$ to each side.

= ⬚   Add.

= ⬚   Simplify.

= ⬚   Factor.

The last expression is the right side of the equation to be proved, where n has been replaced by $k + 1$. Thus, the equation is true for $n = k + 1$.

This proves that $1 + 3 + 5 + \ldots + (2n - 1) = n^2$ is true for all positive integers n.

Your Turn

a. Prove that $2 + 4 + 6 + 8 + \ldots + 2n = n(n + 1)$.

b. Prove that $10^n - 1$ is divisible by 9 for all positive integers n.

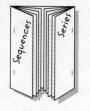

FOLDABLES

ORGANIZE IT

On the tab for Lesson 11-8, write a real-world statement. Then find a counterexample to your statement.

EXAMPLE   Counterexample

② **Find a counterexample for the formula that $n^2 + n + 5$ is always a prime number for any positive integer n.**

Check the first few positive integers.

n	Formula	Prime?
1		
2		
3		
4		

The value $n =$ ☐ is a counterexample for the formula.

HOMEWORK ASSIGNMENT

Page(s): _____

Exercises: _____

Your Turn Find a counterexample for the formula that $2n - 1$ is always a prime number for any positive integer n.

FOLDABLES™	**VOCABULARY PUZZLEMAKER**	**BUILD YOUR VOCABULARY**
Use your **Chapter 11 Foldable** to help you study for your chapter test.	To make a crossword puzzle, word search, or jumble puzzle of the vocabulary words in Chapter 11, go to: www.glencoe.com/sec/math/t_resources/free/index.php.	You can use your completed **Vocabulary Builder** (pages 246–247) to help you solve the puzzle.

11-1
Arithmetic Sequences

1. Find the next four terms of the arithmetic sequence 3, 6, 9, 12, ...

2. Find the first five terms of the arithmetic sequence in which $a_1 = 2$ and $d = 9$.

3. Write an equation for the nth term of the arithmetic sequence 10, 6, 2, −2, ...

11-2
Arithmetic Series

4. Find S_n for the arithmetic series in which $a_1 = -6$, $n = 18$, and $d = 2$.

5. Find the sum of the arithmetic series 30 + 25 + . . . + (−10).

Find the sum of each arithmetic series.

6. $\sum_{j=3}^{9} (6 - j)$

7. $\sum_{k=10}^{25} (2k + 1)$

11-3
Geometric Sequences

8. In the sequence 5, 8, 11, 14, 17, 20, the numbers 8, 11, 14, and

   17 are [   ] between 5 and 20.

9. In the sequence 12, 4, $\frac{4}{3}$, $\frac{4}{9}$, $\frac{4}{27}$, the numbers 4, $\frac{4}{3}$, and $\frac{4}{9}$ are

   [   ] between 12 and $\frac{4}{27}$.

10. Find three geometric means between 4 and 324.

    [   ]

11-4
Geometric Series

11. Consider the formula $S_n = \frac{a_1(1 - r^n)}{1 - r}$. Suppose that you

    want to use the formula to evaluate the sum $\sum_{n=1}^{6} 8(-2)^{n-1}$.

    Indicate the values you would substitute into the formula in

    order to find S_n.

    $n =$ [   ] $a_1 =$ [   ] $r =$ [   ] $r^n =$ [   ]

12. Find the sum of a geometric series for which $a_1 = 5$, $n = 9$, and

    $r = 3$.

    [   ]

11-5
Infinite Geometric Series

13. Consider the formula $S = \frac{a_1}{1 - r}$. For what values of r does an

    infinite geometric sequence have a sum? [   ]

14. For the geometric series $\frac{2}{3} + \frac{2}{9} + \frac{2}{27} + \cdots$, give the values of a_1

    and r. Then state whether the sum of the series exists.

    [   ]

11-6

Recursion and Special Sequences

15. Find the first five terms of the sequence in which $a_1 = -3$ and $a_{n+1} = 2a_n + 5$.

16. Find the first three iterates of $f(x) = 3x - 2$ for an initial value of $x_0 = 4$.

11-7

The Binomial Theorem

Consider the expansion of $(w + z)^5$.

17. How many terms does this expansion have?

18. In the second term of the expansion, what is the exponent of w?

19. In the fourth term of the expansion, what is the exponent of w?

20. What is the last term of this expansion?

11-8

Proof and Mathematical Induction

Suppose that you wanted to prove that the following statement is true for all positive integers.

$$3 + 6 + 9 + \ldots + 3n = \frac{3n(n + 1)}{2}$$

21. Which statement shows that the statement is true for $n = 1$?

| i. $3 = \dfrac{3 \cdot 2 + 1}{2}$ | ii. $3 = \dfrac{3 \cdot 1 \cdot 2}{2}$ | iii. $3 = \dfrac{3 + 1 + 2}{2}$ |

ARE YOU READY FOR THE CHAPTER TEST?

Visit **algebra2.com** to access your textbook, more examples, self-check quizzes, and practice tests to help you study the concepts in Chapter 11.

Check the one that applies. Suggestions to help you study are given with each item.

☐ **I completed the review of all or most lessons without using my notes or asking for help.**

- You are probably ready for the Chapter Test.

- You may want to take the Chapter 11 Practice Test on page 627 of your textbook as a final check.

☐ **I used my Foldable or Study Notebook to complete the review of all or most lessons.**

- You should complete the Chapter 11 Study Guide and Review on pages 622–626 of your textbook.

- If you are unsure of any concepts or skills, refer back to the specific lesson(s).

- You may also want to take the Chapter 11 Practice Test on page 627.

☐ **I asked for help from someone else to complete the review of all or most lessons.**

- You should review the examples and concepts in your Study Notebook and Chapter 11 Foldable.

- Then complete the Chapter 11 Study Guide and Review on pages 622–626 of your textbook.

- If you are unsure of any concepts or skills, refer back to the specific lesson(s).

- You may also want to take the Chapter 11 Practice Test on page 627.

Student Signature   Parent/Guardian Signature

Teacher Signature

Probability and Statistics

 Use the instructions below to make a Foldable to help you organize your notes as you study the chapter. You will see Foldable reminders in the margin of this Interactive Study Notebook to help you in taking notes.

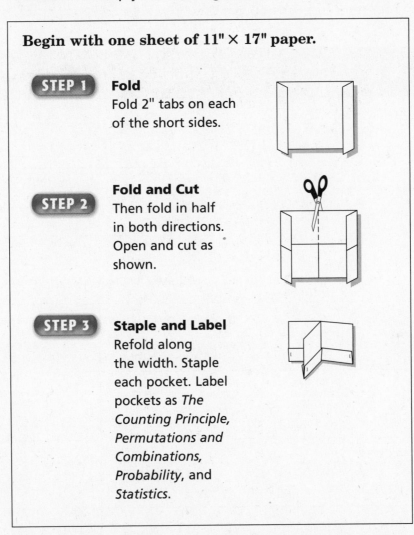

Begin with one sheet of 11" × 17" paper.

STEP 1 **Fold**
Fold 2" tabs on each of the short sides.

STEP 2 **Fold and Cut**
Then fold in half in both directions. Open and cut as shown.

STEP 3 **Staple and Label**
Refold along the width. Staple each pocket. Label pockets as *The Counting Principle, Permutations and Combinations, Probability,* and *Statistics.*

NOTE-TAKING TIP: When you take notes, look for written real-world examples in your everyday life. Comment on how writers use statistics to prove or disprove points of view and discuss the ethical responsibilities writers have when using statistics.

CHAPTER 12

BUILD YOUR VOCABULARY

This is an alphabetical list of new vocabulary terms you will learn in Chapter 12. As you complete the study notes for the chapter, you will see Build Your Vocabulary reminders to complete each term's definition or description on these pages. Remember to add the textbook page number in the second column for reference when you study.

Vocabulary Term	Found on Page	Definition	Description or Example
binomial experiment			
combination			
compound event			
dependent and independent events			
inclusive events [ihn-KLOO-sihv]			
margin of sampling error			
measure of central tendency			
measure of variation			
mutually exclusive events [MYOO-chuh-lee]			

© Glencoe/McGraw-Hill

Vocabulary Term	Found on Page	Definition	Description or Example
normal distribution			
odds			
permutation [puhr-myoo-TAY-shuhn]			
probability			
probability distribution			
random variable			
relative-frequency histogram			
sample space			
skewed distribution [SKYOOD]			
standard deviation			
variance [VEHR-ee-uhn(t)s]			

The Counting Principle

WHAT YOU'LL LEARN

- Solve problems involving independent events.
- Solve problems involving dependent events.

The set of all possible [ ] is called the **sample space**.

An **event** consists of one or more outcomes of a trial. Events that do not affect each other are called **independent events**.

EXAMPLE Fundamental Counting Principle

1. For their vacation, the Murray family is choosing a trip to the beach or to the mountains. They can select their transportation from a car, plane, or train. How many different ways can they select a destination followed by a means of transportation?

There are [ ] · [ ] or [ ] ways to choose a trip.

KEY CONCEPT

Fundamental Counting Principle If event M can occur in m ways and is followed by event N that can occur in n ways, then event M followed by event N can occur in $m \cdot n$ ways.

Your Turn A pizza place offers customers a choice of American, mozzarella, Swiss, feta, or provolone cheese with one topping chosen from pepperoni, mushrooms, or sausage. How many different combinations of cheese and toppings are there?

EXAMPLE More than Two Independent Events

2. COMMUNICATION How many codes are possible if an answering machine requires a 2-digit code to retrieve messages?

The choice of any digit does not affect the other digit, so the choices of digits are independent events.

There are [ ] possible choices for the first digit and

[ ] possible choices for the second digit.

So, there are [ ] · [ ] or [ ] possible different codes.

Your Turn Many automated teller machines (ATM) require a 4-digit code to access an account. How many codes are possible?

ORGANIZE IT

Write and solve your own examples of independent and dependent events. Place your work in The Counting Principle pocket.

BUILD YOUR VOCABULARY (page 270)

With **dependent events**, the outcome of one event

[ ] affect the outcome of another event.

EXAMPLE Dependent Events

3 **How many different schedules could a student have who is planning to take 4 different classes? Assume each class is offered each period.**

The choices of which class to schedule each period are dependent events.

There are 4 classes that can be taken during the first period. That leaves 3 classes for the second period, 2 classes for the third period, and so on.

Period	1st	2nd	3rd	4th
Number of Choices				

There are [ ] · [ ] · [ ] · [ ] or [ ] different

schedules for a student who is taking 4 classes.

Your Turn How many different schedules could a student have who is planning to take 5 different classes?

© Glencoe/McGraw-Hill

HOMEWORK ASSIGNMENT

Page(s):

Exercises:

Permutations and Combinations

WHAT YOU'LL LEARN

- Solve problems involving linear permutations.

- Solve problems involving combinations.

KEY CONCEPTS

Permutations The number of permutations of n distinct objects taken r at a time is given by $P(n, r) = \dfrac{n!}{(n-r)!}$.

Permutations with Repetitions The number of permutations of n objects of which p are alike and q are alike is $\dfrac{n!}{p!q!}$.

> **BUILD YOUR VOCABULARY** (page 271)
>
> When a group of objects or people are arranged in a certain order, the arrangement is called a **permutation**.

EXAMPLE Permutation

① **Eight people enter the BestPic contest. How many ways can blue, red, and green ribbons be awarded?**

Since each winner will receive a different ribbon, order is important. You must find the number of permutations of 8 things taken 3 at a time.

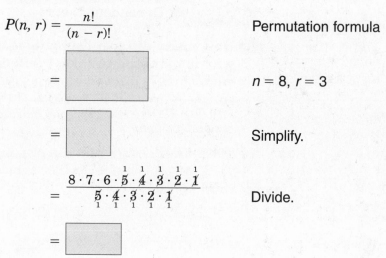

$P(n, r) = \dfrac{n!}{(n-r)!}$ Permutation formula

$=$ $n = 8, r = 3$

$=$ Simplify.

$= \dfrac{8 \cdot 7 \cdot 6 \cdot \overset{1}{\cancel{5}} \cdot \overset{1}{\cancel{4}} \cdot \overset{1}{\cancel{3}} \cdot \overset{1}{\cancel{2}} \cdot \overset{1}{\cancel{1}}}{\underset{1}{\cancel{5}} \cdot \underset{1}{\cancel{4}} \cdot \underset{1}{\cancel{3}} \cdot \underset{1}{\cancel{2}} \cdot \underset{1}{\cancel{1}}}$ Divide.

$=$

EXAMPLE Permutation with Repetition

② **How many different ways can the letters of the word BANANA be arranged?**

The second, fourth, and sixth letters are each A.

The third and fifth letters are each N.

Find the number of permutations of ☐ letters of which

☐ of one letter and ☐ of another letter are the same.

$\dfrac{6!}{3!2!} =$ 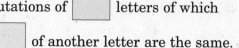 or ☐

Your Turn

a. Ten people are competing in a swim race where 4 ribbons will be given. How many ways can blue, red, green, and yellow ribbons be awarded?

b. How many different ways can the letters of the word ALGEBRA be arranged?

BUILD YOUR VOCABULARY (page 270)

An arrangement or selection of objects in which order is *not* important is called a **combination**.

EXAMPLE Multiple Events

3 **Six cards are drawn from a standard deck of cards. How many hands consist of two hearts and four spades?**

Multiply the number of ways to select two hearts and the number of ways to select four spades. Only the cards in the hand matter, not the order in which they were drawn, so use combinations.

$C(13, 2) \cdot C(13, 4)$

$$= \frac{13!}{(13 - 2)!2!} \cdot \frac{13!}{(13 - 4)!4!}$$ Combination formula

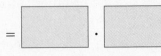 = [ ] · [ ]   Subtract.

= [ ] · [ ] or [ ]   Simplify.

Your Turn

a. Six friends at a party decide that three of them will go to pick up a movie. How many ways can they choose three people to go?

b. Thirteen cards are drawn from a standard deck of cards. How many hands consist of six hearts and seven diamonds?

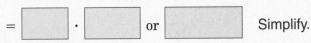

KEY CONCEPTS

Combinations The number of combinations of n distinct objects taken r at a time is given by $C(n, r) = \frac{n!}{(n - r)!r!}$.

FOLDABLES Write a real-world example of a permutation and a combination. Place your work in the Permutations and Combinations pocket.

HOMEWORK ASSIGNMENT

Page(s): _____
Exercises: _____

© Glencoe/McGraw-Hill

Probability

WHAT YOU'LL LEARN

- Find the probability and odds of events.
- Create and use graphs of probability distributions.

BUILD YOUR VOCABULARY (page 271)

The **probability** of an event is a [ ] that measures the chances of the event occurring.

EXAMPLE Probability

1 **When three coins are tossed, what is the probability that all three are heads?**

You can use a tree diagram to find the sample space.

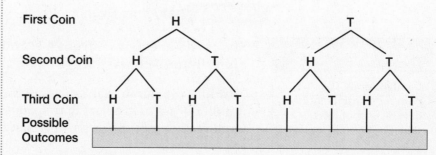

There are [ ] possible outcomes. Only one of these outcomes is HHH, so $s =$ [ ]. The other outcomes are failures, so $f =$ [ ].

$P(\text{3 Heads}) = \dfrac{s}{s + f}$ Probability formula

$= \dfrac{\boxed{}}{\boxed{} + \boxed{}}$ or $\boxed{}$

The probability of tossing three heads is [ ].

Your Turn When three coins are tossed, what is the probability that exactly two are heads?

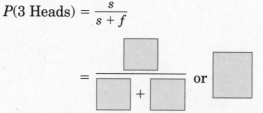

KEY CONCEPT

Probability of Success and Failure If an event can succeed in s ways and fail in f ways, then the probabilities of success, $P(S)$, and of failure, $P(F)$, are as follows.

- $P(S) = \dfrac{s}{s + f}$
- $P(F) = \dfrac{f}{s + f}$

© Glencoe/McGraw-Hill

EXAMPLE Odds

2 LIFE EXPECTANCY According to the U.S. National Center for Health Statistics, the chances of a male born in 1990 living to be at least 65 years of age are about 3 in 4. For females, the chances are about 17 in 20.

KEY CONCEPT

Odds The odds that an event will occur can be expressed as the ratio of the number of ways it can succeed to the number of ways it can fail. If an event can succeed in *s* ways and fail in *f* ways, then the odds of success and of failure are as follows.

• Odds of success = *s*:*f*

• Odds of failure = *f*:*s*

a. What are the odds that a male born in 1990 will die before age 65?

Three out of four males will live to be at least 65, so the number of successes (living to 65) is [ ]. The number of failures is [ ] – [ ] or [ ].

Odds of a male dying before age 65

= *f*:*s*   Odds formula

= [ ]   *f* = 1, *s* = 3

b. What are the odds that a female born in 1990 will die before age 65?

Seventeen out of twenty females will live to be at least 65, so the number of successes in this case is [ ]. The number of failures is [ ] – [ ] or [ ].

Odds of a female dying before age 65

= *f*:*s*   Odds formula

= [ ]   *f* = 3, *s* = 17

Your Turn The chances of a male born in 1980 to live to be at least 65 years of age are about 7 in 10. For females, the chances are about 21 in 25.

a. What are the odds that a male born in 1980 will live to age 65?

b. What are the odds that a female born in 1980 will live to age 65?

BUILD YOUR VOCABULARY (page 271)

A **random variable** is a variable whose value is the numerical outcome of a [_____] event.

A **probability distribution** for a particular random variable is a function that maps the [_____] to the [_____] of the outcomes in the sample space.

To help visualize a [_____] distribution, you can use a table of probabilities or a graph, called a **relative-frequency histogram**.

EXAMPLE Probability Distribution

3 Suppose two dice are rolled. The table and the relative-frequency histogram show the distribution of the sum of the numbers rolled.

S = Sum	2	3	4	5	6	7	8	9	10	11	12
Probability	$\frac{1}{36}$	$\frac{1}{18}$	$\frac{1}{12}$	$\frac{1}{9}$	$\frac{5}{36}$	$\frac{1}{6}$	$\frac{5}{36}$	$\frac{1}{9}$	$\frac{1}{12}$	$\frac{1}{18}$	$\frac{1}{36}$

a. Use the graph to determine which outcomes are least likely. What are their probabilities?

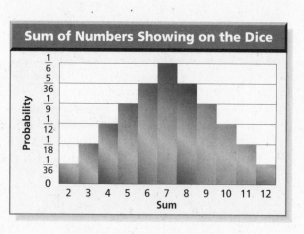

The least probability is [_____]. The least likely outcomes are sums of [_____] and [_____].

b. What are the odds of rolling a sum of 5?

STEP 1
Identify s and f.

$$P(\text{rolling a 5}) = \frac{s}{s+f} \quad s = 1, f = 8$$

$$= \boxed{}$$

STEP 2
Find the odds.

$$\text{Odds} = s{:}f.$$

$$= \boxed{}$$

So, the odds of rolling a sum of 5 are $\boxed{}$.

Your Turn Suppose two dice are rolled. The table and the relative frequency histogram show the distribution of the sum of the numbers rolled.

S = Sum	2	3	4	5	6	7	8	9	10	11	12
Probability	$\frac{1}{36}$	$\frac{1}{18}$	$\frac{1}{12}$	$\frac{1}{9}$	$\frac{5}{36}$	$\frac{1}{6}$	$\frac{5}{36}$	$\frac{1}{9}$	$\frac{1}{12}$	$\frac{1}{18}$	$\frac{1}{36}$

a. Use the graph to determine which outcomes are the second most likely. What are their probabilities?

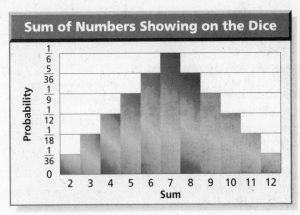

Sum of Numbers Showing on the Dice

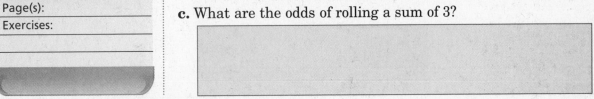

b. Use the table to find $P(S = 4)$. What other sum has the same probability?

c. What are the odds of rolling a sum of 3?

HOMEWORK ASSIGNMENT

Page(s):

Exercises:

Multiplying Probabilities

WHAT YOU'LL LEARN

- Find the probability of two independent events.

- Find the probability of two dependent events.

1 Gerardo has 9 dimes and 7 pennies in his pocket. He randomly selects one coin, looks at it, and replaces it. He then randomly selects another coin. What is the probability that both of the coins he selects are dimes?

P(both dimes)

$= P(\text{dime}) \cdot P(\text{dime})$ Probability of independent events

$=$ [ ] $\cdot$ [ ]   Substitute.

$=$ [ ]   Multiply.

The probability is [ ] or about [ ] %.

KEY CONCEPT

Probability of Two Independent Events If two events, A and B, are independent, then the probability of both events occurring is $P(A \text{ and } B) = P(A) \cdot P(B)$.

Your Turn

a. Gerardo has 9 dimes and 7 pennies in his pocket. He randomly selects one coin, looks at it, and replaces it. He then randomly selects another coin. What is the probability that both of the coins he selects are pennies?

b. When three dice are rolled, what is the probability that one die is a multiple of 3, one die shows an even number, and one die shows a 5?

© Glencoe/McGraw-Hill

KEY CONCEPT

Probability of Two Dependent Events If two events, *A* and *B*, are dependent, then the probability of both events occurring is
$P(A \text{ and } B) =$
$P(A) \cdot P(B \text{ following } A)$.

FOLDABLES

ORGANIZE IT

Explain the difference between the probability of two independent events and the probability of two dependent events. Place your work in the Probability pocket.

HOMEWORK ASSIGNMENT

Page(s):

Exercises:

EXAMPLE Two Dependent Events

2 The host of a game show draws chips from a bag to determine the prizes for which contestants will play. Of the 20 chips, 11 show *computer*, 8 show *trip*, and 1 shows *truck*. If the host draws the chips at random and does not replace them, find each probability.

a. a computer, then a truck

$P(C \text{ then } T)$

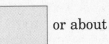

Dependent events

$=$  or

After the first chip is drawn, there are 19 left.

The probability is or about .

b. two trips

$P(T \text{ then } T) = P(T) \cdot P(T \text{ following } T)$

Dependent events

$=$ $\cdot$ or

If the first chip shows *trip*, then 7 of the remaining 19 show *trip*.

The probability is 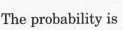 or about .

Your Turn The host of a game show draws chips from a bag to determine the prizes for which contestants will play. Of the 20 chips in the bag, 11 show *computer*, 8 show *trip*, and 1 shows *boat*. If the host draws the chips at random and does not replace them, find each probability.

a. a boat, then a trip

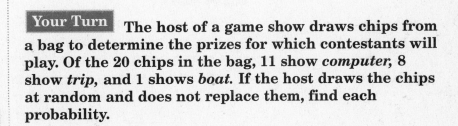

b. two computers

Adding Probabilities

WHAT YOU'LL LEARN

- Find the probability of mutually exclusive events.
- Find the probability of inclusive events.

BUILD YOUR VOCABULARY (page 270)

An event that consists of two or more [   ] events is called a **compound event**.

If two events cannot occur at the same [   ], they are called **mutually exclusive events**.

KEY CONCEPT

Probability of Mutually Exclusive Events If two events, *A* and *B*, are mutually exclusive, then the probability that *A* or *B* occurs is the sum of their probabilities.

EXAMPLE Two Mutually Exclusive Events

① **Sylvia has a stack of playing cards consisting of 10 hearts, 8 spades, and 7 clubs. If she selects a card at random from this stack, what is the probability that it is a heart or club?**

These are mutually exclusive events since the card cannot be both a heart and a club.

P(heart or club)

= *P*(H) + *P*(C)   Mutually exclusive events

= [   ] + [   ] or [   ]   Substitute and add.

The probability that Sylvia selects a heart or a club is [   ].

Your Turn

a. Sylvia has a stack of playing cards consisting of 10 hearts, 8 spades, and 7 clubs. If she selects a card at random from this stack, what is the probability that it is a spade or a club?

b. The Book Club makes a list of 9 mysteries and 3 romance books they want to read. They plan to select 3 titles at random to read this semester. What is the probability that at least two of the books they select are romances?

BUILD YOUR VOCABULARY (page 270)

If two events are not [_____], they are called **inclusive events**.

EXAMPLE Inclusive Events

KEY CONCEPT

Probability of Inclusive Events If two events, *A* and *B*, are inclusive, then the probability that *A* or *B* occurs is the sum of their probabilities decreased by the probability of both occurring.

 There are 2400 subscribers to an Internet service provider. Of these, 1200 own Brand A computers, 500 own Brand B, and 100 own both A and B. What is the probability that a subscriber selected at random owns either Brand A or Brand B?

Since some subscribers own both *A* and *B*, the events are inclusive.

$P(A) = $ [____] $P(B) = $ [____] $P(\text{both}) = $ [____]

$P(A \text{ or } B) = P(A) + P(B) - P(A \text{ and } B)$

$= $ [____] $+$ [____] $-$ [____]

$= $ [____]   Substitute and simplify.

The probability that a subscriber owns either A or B is

[____] .

Your Turn There are 200 students taking Calculus, 500 taking Spanish, and 100 taking both. There are 1000 students in the school. What is the probability that a student selected at random is taking Calculus or Spanish?

HOMEWORK ASSIGNMENT

Page(s): _____

Exercises: _____

Statistical Measures

WHAT YOU'LL LEARN

- Use measures of central tendency to represent a set of data.

- Find measures of variation for a set of data.

BUILD YOUR VOCABULARY (pages 270–271)

A number that describes a set of data is called a **measure of central tendency** because it represents the [＿＿＿] or middle of the data.

Measures of variation or dispersion measure how [＿＿＿] or [＿＿＿] a set of data is.

The **standard deviation** σ is the [＿＿＿] of the **variance**.

EXAMPLE Choose a Measure of Central Tendency

① A new Internet company has 3 employees who are paid $300,000, 10 who are paid $100,000, and 60 who are paid $50,000. Which measure of central tendency best represents the pay at this company?

Since most of the employees are paid $50,000, the higher values are outliers.

Thus, the [＿＿＿] or [＿＿＿] best represents the pay at this company.

Your Turn In a cereal contest, there is 1 Grand Prize of $1,000,000, 10 first prizes of $100, and 50 second prizes of $10.

a. Which measure of central tendency best represents the prizes?

[＿＿＿＿＿＿＿＿＿＿]

b. Which measure of central tendency would advertisers be most likely to use?

[＿＿＿＿＿＿＿＿＿＿]

REVIEW IT

In your own words, what is an outlier? (*Lesson 2-5*)

EXAMPLE Standard Deviation

KEY CONCEPT

Standard Deviation If a set of data consists of the *n* values x_1, x_2, ..., x_n and has mean x, then the standard deviation σ is given by the following formula.

$$\sigma = \sqrt{\frac{(x_1 - \bar{x})^2 + (x_2 - \bar{x})^2 + \cdots + (x_n - \bar{x})^2}{n}}$$

FOLDABLES Suppose you have a small standard of deviation for your test scores. Does this mean that you have been consistent or inconsistent? Place your explanation in the Statistics pocket.

HOMEWORK ASSIGNMENT

Page(s):

Exercises:

2 **RIVERS** This table shows the length in thousands of miles of some of the longest rivers in the world. Find the standard deviation for these data.

River	Length (thousands of miles)
Nile	4.16
Amazon	4.08
Missouri	2.35
Rio Grande	1.90
Danube	1.78

Find the mean. Add the data and divide by the number of items.

$$\bar{x} = \frac{4.16 + 4.08 + 2.35 + 1.90 + 1.78}{5}$$

$$= \boxed{} \text{ thousand miles}$$

Find the variance.

$$\sigma^2 = \frac{(x_1 - \bar{x})^2 + (x_2 - \bar{x})^2 + \cdots + (x_n - \bar{x})^2}{n} \qquad \text{Variance formula}$$

$$\approx \frac{(4.16 - 2.85)^2 + (4.08 - 2.85)^2 + \cdots + (1.78 - 2.85)^2}{5}$$

$$\approx \boxed{} \qquad \text{Simplify.}$$

$$= \boxed{} \text{ thousand miles}$$

Find the standard deviation.

$$\sigma^2 \approx \boxed{} \qquad \text{Take the square root of each side.}$$

$$\sigma \approx \boxed{} \text{ thousand miles}$$

Your Turn A teacher has the following test scores: 100, 4, 76, 85, and 92. Find the standard deviation for these data.

The Normal Distribution

WHAT YOU'LL LEARN

- Determine whether a set of data appears to be normally distributed or skewed.

- Solve problems involving normally distributed data.

BUILD YOUR VOCABULARY (page 271)

A curve that is [   ], often called a *bell curve*, is called a **normal distribution**.

A curve or histogram that in not symmetric represents a **skewed distribution**.

EXAMPLE Classify a Data Distribution

① Determine whether the data {31, 33, 35, 33, 36, 34, 36, 32, 36, 33, 32, 34, 34, 35, 34} appear to be *positively skewed*, *negatively skewed*, or *normally distributed*.

Make a frequency table for the data. Then use the table to make a histogram.

Value	31	32	33	34	35	36
Frequency						

Since the data are somewhat symmetric, this is a

[   ].

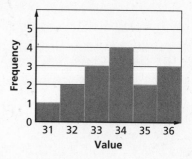

REMEMBER IT

To help remember how skewed distributions are labeled, think about the long tail being in the direction of the skew. For example, a positively skewed distribution has a long tail in the positive direction.

Your Turn Determine whether the data {7, 5, 6, 7, 8, 4, 6, 8, 7, 6, 6, 4} shown in the histogram appear to be *positively skewed*, *negatively skewed*, or *normally distributed*.

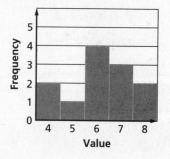

EXAMPLE Normal Distribution

2 Students counted the number of candies in 100 small packages. They found that the number of candies per package was normally distributed with a mean of 23 candies per package and a standard deviation of 1 piece of candy. About how many packages had between 24 and 22 candies?

Draw a normal curve. Label the mean and positive and negative multiples of the standard deviation.

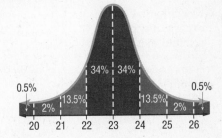

The values of 22 and 24 are [ ] standard deviation *below*

and above the mean, respectively. Therefore, [ ] of the data are located here.

100 · [ ] = [ ] packages   Multiply 100 by 0.68.

About [ ] packages contained between 22 and 24 pieces.

Your Turn Students counted the number of candies in 100 small packages. They found that the number of candies per package was normally distributed with a mean of 23 candies per package and a standard deviation of 1 piece of candy.

a. About how many packages had between 25 and 21 candies?

b. What is the probability that a package selected at random had more than 24 candies?

FOLDABLES

ORGANIZE IT
Describe real-world situations where you would expect the data to be positively skewed, negatively skewed, and normally distributed. Place your explanations in the Statistics pocket.

HOMEWORK ASSIGNMENT
Page(s):
Exercises:

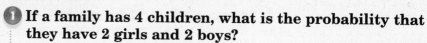
EXAMPLE Binomial Theorem

① **If a family has 4 children, what is the probability that they have 2 girls and 2 boys?**

There are two possible outcomes for the gender of each of their children: boy or girl. The probability of a boy b is ⬚, and the probability of a girl g is ⬚.

$$(b + g)^4 = \boxed{} + \boxed{} + \boxed{} + \boxed{} + \boxed{}$$

The term $6b^2g^2$ represents 2 girls and 2 boys.

$P(\text{2 girls and 2 boys})$

$= 6b^2g^2$

$= 6\left(\boxed{}\right)^2 \left(\boxed{}\right)^2 \qquad b = \boxed{}, g = \boxed{}$

$= \boxed{} \qquad\qquad\qquad\qquad$ Multiply.

The probability of 2 boys and 2 girls is ⬚ or 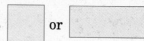 .

Your Turn If a family has 4 children, what is the probability that they have 4 boys?

EXAMPLE Binomial Experiment

KEY CONCEPT

Binomial Experiments
A binomial experiment exists if and only if all of these conditions occur.

- There are exactly two possible outcomes for each trial.

- There is a fixed number of trials.

- The trials are independent.

- The probabilities for each trial are the same.

2 A report said that approximately 1 out of 6 cars sold in a certain year was green. Suppose a salesperson sells 7 cars per week. What is the probability that this salesperson will sell exactly 3 green cars in a week?

The probability that a sold car is green is .

The probability that a sold car is not green is .

There are $C(7, 3)$ ways to choose the three green cars that sell.

$P(3 \text{ green cars}) = C(7, 3) \left(\dfrac{1}{6}\right)^3 \left(\dfrac{5}{6}\right)^4$

If he sells three green cars, he sells four that are not green.

$= \boxed{} \left(\dfrac{1}{6}\right)^3 \left(\dfrac{5}{6}\right)^4$

$C(7, 3) = \dfrac{7!}{4!3!}$

$= \boxed{}$ Simplify.

The probability that he will sell exactly 3 green cars is

$\boxed{}$ or about $\boxed{}$.

Your Turn A report said that approximately 1 out of 6 cars sold in a certain year was green. Suppose a salesperson sells 7 cars per week.

a. What is the probability that this salesperson will sell exactly 4 green cars in a week?

b. What is the probability that this salesperson will sell at least 2 green cars in a week?

HOMEWORK ASSIGNMENT

Page(s):

Exercises:

Sampling and Error

WHAT YOU'LL LEARN

- Determine whether a sample is unbiased.

- Find margins of sampling error.

KEY CONCEPT

Margin of Sampling Error
If the percent of people in a sample responding in a certain way is p and the size of the sample is n, then 95% of the time, the percent of the population responding in that same way will be between $p - ME$ and $p + ME$, where

$$ME = 2\sqrt{\frac{p(1 - p)}{n}}.$$

EXAMPLE Biased and Unbiased Samples

1 State whether surveying people going into an action movie to find out the most popular kind of movie would produce a random sample. Explain.

[   ]; they will most likely think that action movies are the most popular kind of movie.

Your Turn State whether each method would produce a random sample. Explain.

a. surveying people going into a football game to find out the most popular sport

b. surveying every fifth person going into a mall to find out the most popular kind of movie

EXAMPLE Find a Margin of Error

2 In a survey of 100 randomly selected adults, 37% answered "yes" to a particular question. What is the margin of error?

$$ME = 2\sqrt{\frac{p(1 - p)}{n}}$$

Formula for margin of sampling error

$$= \underline{\hspace{3cm}}$$

$p = 37\%$ or 0.37, $n = 100$

$$\approx \underline{\hspace{2cm}} \text{ or } 10\%$$

Use a calculator.

This means that there is a 95% chance that the percent of people in the whole population who would answer "yes"

is between $37 - 10$ or [   ]% and $37 + 10$ or [   ]%.

Your Turn In a survey of 100 randomly selected adults, 50% answered "no" to a particular question. What is the margin of error?

EXAMPLE Analyze a Margin of Error

3 **HEALTH** In an earlier survey, 30% of the people surveyed said they had smoked cigarettes in the past week. The margin of error was 2%.

a. **What does the 2% indicate about the results?**

There is a [   ] % chance that the percent of people in the population who had smoked cigarettes in the past week

was between [   ] % and [   ] %.

b. **How many people were surveyed?**

$$ME = 2\sqrt{\frac{p(1-p)}{n}}$$ Formula for margin of sampling error

$$0.02 = 2\sqrt{\frac{0.3(1-0.3)}{n}}$$ $ME = 0.02$, $p = 0.3$

[   ] = [   ]   Divide by 2.

$$0.0001 = \frac{0.21}{n}$$ Square each side.

$$n = \frac{0.21}{0.0001}$$ Multiply by n and divide by 0.0001.

$$n = [\quad]$$ Use a calculator.

Your Turn In an earlier survey, 25% of the people surveyed said they had exercised in the past week. The margin of error was 2%.

a. What does the 2% indicate about the results?

b. How many people were surveyed?

FOLDABLES

ORGANIZE IT

Write your own example of a biased and an unbiased survey. Place your work in the Statistics pocket.

HOMEWORK ASSIGNMENT

Page(s):

Exercises:

STUDY GUIDE

FOLDABLES™	**VOCABULARY PUZZLEMAKER**	**BUILD YOUR VOCABULARY**
Use your **Chapter 12 Foldable** to help you study for your chapter test.	To make a crossword puzzle, word search, or jumble puzzle of the vocabulary words in Chapter 12, go to: www.glencoe.com/sec/math/ t_resources/free/index.php.	You can use your completed **Vocabulary Builder** (pages 270–271) to help you solve the puzzle.

12-1

The Counting Principle

A jar contains 6 red marbles, 4 blue marbles, and 3 yellow marbles. Indicate whether the events described are *dependent* or *independent*.

1. A marble is drawn out of the jar and is not replaced. A second marble is drawn.

2. A marble is drawn out of the jar and is put back in. The jar is shaken. A second marble is drawn.

3. A man owns two suits, ten ties, and eight shirts. How many different outfits can he wear if each is made up of a suit, a tie, and a shirt?

12-2

Permutations and Combinations

4. Indicate whether arranging five pictures in a row on a wall involves a *permutation* or a *combination*.

Evaluate each expression.

5. $P(5, 3)$

6. $C(7, 2)$

12-3

Probability

A weather forecast says that the chance of rain tomorrow is 40%.

7. Write the probability that it will rain tomorrow as a fraction in lowest terms.

8. What are the odds in favor of rain?

9. Balls are numbered 1 through 15. Find the probability that a ball drawn at random will show a number less than 4. Then find the odds that a number less than 4 is drawn.

12-4

Multiplying Probabilities

A bag contains 4 yellow balls, 5 red balls, 1 white ball, and 2 black balls. A ball is drawn from the bag and is not replaced. A second ball is drawn.

10. Tell which formula you would use to find the probability that the first ball is yellow and the second ball is black.

   a. $P(Y \text{ and } B) = \dfrac{P(Y)}{P(Y) + P(B)}$ **b.** $P(Y \text{ and } B) = P(Y) \cdot P(B)$

   c. $P(Y \text{ and } B) = P(Y) \cdot P(B \text{ following } Y)$

11. Which equation shows the correct calculation of this probability?

   a. $\dfrac{1}{3} + \dfrac{2}{11} = \dfrac{17}{33}$ **b.** $\dfrac{1}{3} \cdot \dfrac{2}{11} = \dfrac{2}{33}$

   c. $\dfrac{1}{3} + \dfrac{1}{6} = \dfrac{1}{2}$ **d.** $\dfrac{1}{3} \cdot \dfrac{1}{6} = \dfrac{1}{18}$

12. A pair of dice is thrown. What is the probability that both dice show a number greater than 5?

12-5
Adding Probabilities

Marla took a quiz on this lesson that contained the following problem. Her solution is shown.

Each of the integers from 1 through 25 is written on a slip of paper and placed in an envelope. If one slip is drawn at random, what is the probability that it is odd or a multiple of 5?

$P(\text{odd}) = \dfrac{13}{25}$ $P(\text{multiple of 5}) = \dfrac{5}{25}$ or $\dfrac{1}{5}$

$P(\text{odd or multiple of 5}) = P(\text{odd}) + P(\text{multiple of 5})$

$$= \dfrac{13}{25} + \dfrac{5}{25} = \dfrac{18}{25}$$

13. Why is Marla's work incorrect?

14. Show the corrected work.

15. A card is drawn from a standard deck of 52 playing cards. What is the probability that an ace or a black card is drawn?

12-6
Statistical Measures

Consider the data set {25, 31, 49, 52, 68, 79, 105}.

16. Find the variance to the nearest tenth.

17. Find the standard deviation to the nearest tenth.

12-7
The Normal Distribution

Indicate whether each of the following statements is *true* or *false*.

18. In a continuous probability distribution, there is a finite number of possible outcomes.

19. Every normal distribution can be represented by a bell curve.

12-8
Binomial Experiments

Indicate whether each of the following is a *binomial experiment* or *not a binomial experiment*. If the experiment is not a binomial experiment, explain why.

20. A fair coin is tossed 10 times and "heads" or "tails" is recorded each time.

21. A pair of dice is thrown 5 times and the sum of the numbers that come up is recorded each time.

Find each probability if a coin is tossed four times.

22. P(exactly three heads)

23. P(exactly four heads)

12-9
Sampling and Error

25. In a survey of 200 people, 36% voted in the last presidential election. Find the margin of sampling error.

ARE YOU READY FOR THE CHAPTER TEST?

Visit **algebra2.com** to access your textbook, more examples, self-check quizzes, and practice tests to help you study the concepts in Chapter 12.

Check the one that applies. Suggestions to help you study are given with each item.

☐ **I completed the review of all or most lessons without using my notes or asking for help.**

- You are probably ready for the Chapter Test.

- You may want to take the Chapter 12 Practice Test on page 693 of your textbook as a final check.

☐ **I used my Foldable or Study Notebook to complete the review of all or most lessons.**

- You should complete the Chapter 12 Study Guide and Review on pages 687–692 of your textbook.

- If you are unsure of any concepts or skills, refer back to the specific lesson(s).

- You may also want to take the Chapter 12 Practice Test on page 693.

☐ **I asked for help from someone else to complete the review of all or most lessons.**

- You should review the examples and concepts in your Study Notebook and Chapter 12 Foldable.

- Then complete the Chapter 12 Study Guide and Review on pages 687–692 of your textbook.

- If you are unsure of any concepts or skills, refer back to the specific lesson(s).

- You may also want to take the Chapter 12 Practice Test on page 693.

Student Signature	Parent/Guardian Signature

Teacher Signature

Trigonometric Functions

 Use the instructions below to make a Foldable to help you organize your notes as you study the chapter. You will see Foldable reminders in the margin of this Interactive Study Notebook to help you in taking notes.

Begin with one sheet of construction paper and two pieces of grid paper.

STEP 1 **Fold and Cut**
Stack and fold on the diagonal. Cut to form a triangular stack.

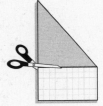

STEP 2 **Staple and Label**
Staple edge to form a booklet.

 NOTE-TAKING TIP: When you take notes, include visuals. Clearly label the visuals and write captions when needed.

BUILD YOUR VOCABULARY

This is an alphabetical list of new vocabulary terms you will learn in Chapter 13. As you complete the study notes for the chapter, you will see Build Your Vocabulary reminders to complete each term's definition or description on these pages. Remember to add the textbook page number in the second column for reference when you study.

Vocabulary Term	Found on Page	Definition	Description or Example
angle of depression or elevation			
Arccosine function [AHRK-KOH-SYN]			
Arcsine function [AHRK-SYN]			
Arctangent function [AHRK-TAN-juhnt]			
cosecant [KOH-SEE-KANT]			
cosine			
coterminal angles			
cotangent			
Law of Cosines			

Vocabulary Term	Found on Page	Definition	Description or Example
Law of Sines			
period			
principal values			
quadrantal angles [kwah-DRAN-tuhl]			
radian [RAY-dee-uhn]			
reference angle			
secant			
sine			
standard position			
tangent			
trigonometry [TRIH-guh-NAH-muh-tree]			

13–1 Right Triangle Trigonometry

WHAT YOU'LL LEARN

- Find values of trigonometric functions for acute angles.

- Solve problems involving right angles.

BUILD YOUR VOCABULARY (pages 298–299)

Trigonometry is the study of the relationships among the

[ ] and [ ] of a right triangle.

EXAMPLE Find Trigonometric Values

1 Find the value of the six trigonometric functions for angle *G*.

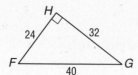

KEY CONCEPT

Trigonometric Functions

$\sin\theta = \dfrac{\text{opp}}{\text{hyp}}$ $\cos\theta = \dfrac{\text{adj}}{\text{hyp}}$

$\tan\theta = \dfrac{\text{opp}}{\text{adj}}$ $\csc\theta = \dfrac{\text{hyp}}{\text{opp}}$

$\sec\theta = \dfrac{\text{hyp}}{\text{adj}}$ $\cot\theta = \dfrac{\text{adj}}{\text{opp}}$

$\csc\theta = \dfrac{1}{\sin\theta}$

$\sec\theta = \dfrac{1}{\cos\theta}$

$\cot\theta = \dfrac{1}{\tan\theta}$

For this triangle, the leg opposite $\angle G$ is [ ] and the leg adjacent to $\angle G$ is [ ]. The hypotenuse is [ ].

Use opp = [ ], adj = [ ], and hyp = [ ] to write each trigonometric ratio.

$\sin G = \dfrac{\text{opp}}{\text{hyp}} =$ [ ] $\cos G = \dfrac{\text{adj}}{\text{hyp}} =$ [ ]

$\tan G = \dfrac{\text{opp}}{\text{adj}} =$ [ ] $\csc G = \dfrac{\text{hyp}}{\text{opp}} =$ [ ]

$\sec G = \dfrac{\text{hyp}}{\text{adj}} =$ [ ] $\cot G = \dfrac{\text{adj}}{\text{opp}} =$ [ ]

Your Turn Find the value of the six trigonometric functions for angle *A*.

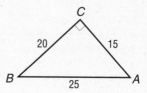

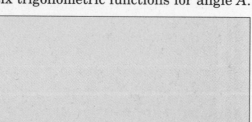

EXAMPLE Find a Missing Side Length of a Right Triangle

2 Write an equation involving sin, cos, or tan that can be used to find the value of x. Then solve the equation. Round to the nearest tenth.

The measure of the hypotenuse is 12.

The side with the missing length is *opposite* the angle measuring 60°. The trigonometric function relating the opposite side of a right triangle and the hypotenuse is the sine function.

$\sin \theta = \dfrac{\text{opp}}{\text{hyp}}$ Sine ratio

$\sin \boxed{} = \boxed{}$ Replace θ with 60°, *opp* with *x*, and *hyp* with 12.

$\boxed{} = \boxed{}$ $\sin 60° = \dfrac{\sqrt{3}}{2}$

$\boxed{} = x$ Multiply each side by 12.

The value of x is $\boxed{}$ or about $\boxed{}$.

Your Turn Write an equation involving sin, cos, or tan that can be used to find the value of x. Then solve the equation. Round to the nearest tenth.

BUILD YOUR VOCABULARY (page 298)

The angle between a $\boxed{}$ line and the line of sight from the observer to an object at a $\boxed{}$ level is called the **angle of elevation**. The angle between a $\boxed{}$ line and the line of sight from the observer to an object at a $\boxed{}$ level is called the **angle of depression**.

FOLDABLES

ORGANIZE IT
On the first page of your Trigonometric Functions booklet, write each of the six trigonometric ratios introduced in this lesson.

EXAMPLE Use an Angle of Elevation

3 SKIING A run has an angle of elevation of 15.7° and a vertical drop of 1800 feet. Estimate the length of this run.

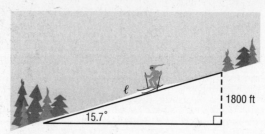

Let ℓ represent the length of the run. Write an equation using a trigonometric function that involves the ratio of ℓ and 1800.

sin 15.7° = ⬚ $\sin \theta = \dfrac{\text{opp}}{\text{hyp}}$

ℓ = ⬚   Solve for ℓ.

$\ell \approx$ ⬚   Use a calculator.

The length of the run is about ⬚ feet.

Your Turn A run has an angle of elevation of 23° and a vertical drop of 1000 feet. Estimate the length of this run.

HOMEWORK ASSIGNMENT

Page(s):
Exercises:

WHAT YOU'LL LEARN

- Change radian measure to degree measure and vice versa.

- Identify coterminal angles.

BUILD YOUR VOCABULARY (page 299)

An angle positioned so that its [ ] is at the origin and its initial side is along the positive *x*-axis is said to be in **standard position**.

One **radian** is the measure of an angle θ in [ ] position whose rays intercept an arc of length 1 unit on the unit circle.

When two angles in standard position have the same [ ] sides, they are called **coterminal angles**.

FOLDABLES

ORGANIZE IT

Use the second page of your Trigonometric Functions booklet. Define and give an example of each new Vocabulary Builder term from the lesson.

EXAMPLE Draw an Angle in Standard Position

1 **Draw the angle −45° in standard position.**

The angle is negative. Draw the terminal side [ ] clockwise from the [ ] *x*-axis.

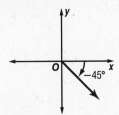

Your Turn Draw the angle 225° in standard position.

EXAMPLE Convert Between Degree and Radian Measure

② Rewrite the degree measure in radians and the radian measure in degrees.

a. **30°**

$$30° = 30° \left(\frac{\pi \text{ radians}}{180°} \right)$$

$$= \boxed{} \text{ radians or } \boxed{}$$

b. $-\dfrac{5\pi}{3}$

$$-\frac{5\pi}{3} = \left(-\frac{5\pi}{3} \text{ radians} \right) \left(\frac{180°}{\pi \text{ radians}} \right)$$

$$= \boxed{} \text{ or } \boxed{}$$

Your Turn Rewrite the degree measure in radians and the radian measure in degrees.

a. 45°

b. $\dfrac{\pi}{6}$

EXAMPLE Find Coterminal Angles

③ Find one angle with positive measure and one angle with negative measure coterminal with

a. **210°**

A positive angle is 210° + 360° or .

A negative angle is 210° − 360° or .

b. $\dfrac{7\pi}{3}$

A positive angle is $\dfrac{7\pi}{3} + 2\pi$ or .

A negative angle is $\dfrac{7\pi}{3} - 2(2\pi)$ or .

Your Turn Find one angle with positive measure and one angle with negative measure coterminal with each angle.

a. 150°

b. $\dfrac{\pi}{6}$

Trigonometric Functions of General Angles

KEY CONCEPT

Trigonometric Functions, θ in Standard Position

Let θ be an angle in standard position and let $P(x, y)$ be a point on the terminal side of θ. Using the Pythagorean Theorem, the distance r from the origin to P is given by $r = \sqrt{x^2 + y^2}$. The trigonometric functions of an angle in standard position may be defined as follows.

$\sin \theta = \dfrac{y}{r}$

$\cos \theta = \dfrac{x}{r}$

$\tan \theta = \dfrac{y}{x}, x \neq 0$

$\csc \theta = \dfrac{r}{y}, y \neq 0$

$\sec \theta = \dfrac{r}{x}, x \neq 0$

$\cot \theta = \dfrac{x}{y}, y \neq 0$

EXAMPLE Evaluate Trigonometric Functions for a Given Point

1 **Find the exact values of the six trigonometric functions of θ if the terminal side of θ contains the point (8, −15).**

From the coordinates given, you know that $x = 8$ and $y = -15$. Use the Pythagorean Theorem to find r.

$r = \sqrt{x^2 + y^2}$ Pythagorean Theorem

$= $ Replace x and y.

$= $ or Simplify.

Now use the values of x, y, and r to write the ratios.

$\sin \theta = \dfrac{y}{r}$ $\cos \theta = \dfrac{x}{r}$

$= $ or $= $

$\tan \theta = \dfrac{y}{x}$ $\csc \theta = \dfrac{r}{y}$

$= $ or $= $ or

$\sec \theta = \dfrac{r}{x}$ $\cot \theta = \dfrac{x}{y}$

$= $ $= $ or

Your Turn Find the exact values of the six trigonometric functions of θ if the terminal side of θ contains the point (−3, 4).

BUILD YOUR VOCABULARY (page 299)

If the [_____] side of angle θ lies on one of the axes, θ is called a **quadrantal angle**.

If θ is a nonquadrantal angle in standard position, its

reference angle, θ', is defined as the [_____] angle

formed by the terminal side of θ and the *x*-axis.

WRITE IT

Give an example of an angle in each quadrant. Be sure to label each angle in degrees and radians.

_____

_____

_____

_____

_____

_____

EXAMPLE   Quadrantal Angles

2 **Find the values of the six trigonometric functions for an angle in standard position that measures 180°.**

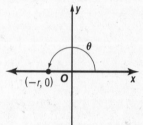

When $\theta = 180$, $x = -r$, and $y = 0$.

$\sin \theta = \dfrac{y}{r}$  $\cos \theta = \dfrac{x}{r}$

$=$  $=$

$\tan \theta = \dfrac{y}{x}$ $\csc \theta = \dfrac{r}{y}$

$=$  $=$

$\sec \theta = \dfrac{r}{x}$ $\cot \theta = \dfrac{x}{y}$

$=$ $=$

Your Turn Find the exact values of the six trigonometric functions for an angle in standard position that measures 90°.

EXAMPLE Find the Reference Angle for a Given Angle

3 Sketch the angle 330°. Then find its reference angle.

Because the terminal side of 330° lies in

quadrant [   ], the reference angle is

[   ] – [   ] or [   ].

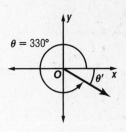

$\theta = 330°$

Your Turn Sketch each angle. Then find its reference angle.

a. 315°

b. $\dfrac{-3\pi}{r}$

FOLDABLES

ORGANIZE IT
On the third page of your Trigonometric Functions booklet, make a sketch of a quadrantal angle and a reference angle.

Trigonometric Functions

HOMEWORK ASSIGNMENT

Page(s): _____
Exercises: _____
_____
_____

EXAMPLE Use a Reference Angle to Find a Trigonometric Value

4 Find the exact value of sin 135°.

Because the terminal side of 135° lies in

Quadrant [   ], the reference angle θ'

is [   ] – [   ] or [   ].

The sine function is [   ] in

Quadrant [   ], so sin 135° = sin 45° or [   ].

$\theta = 135°$
$\theta' = 45°$

Your Turn Find the exact value of each trigonometric function.

a. sin 120° [   ]

b. $\cot \dfrac{11\pi}{6}$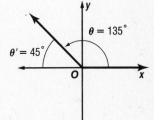

WHAT YOU'LL LEARN

- Solve problems by using the Law of Sines.

- Determine whether a triangle has one, two, or no solutions.

KEY CONCEPTS

Area of a Triangle The area of a triangle is one half the product of the lengths of two sides and the sine of their included angle.

Law of Sines Let $\triangle ABC$ be any triangle with a, b, and c representing the measures of sides opposite angles with measurements A, B, and C, respectively. Then,

$$\frac{\sin A}{a} = \frac{\sin B}{b} = \frac{\sin C}{c}.$$

EXAMPLE Find the Area of a Triangle

1 **Find the area of $\triangle ABC$ to the nearest tenth.**

In this triangle, $b = 6$, $c = 3$, and $A = 25°$. Use the formula $A = \frac{1}{2}bc \sin A$.

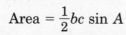

Area $= \frac{1}{2}bc \sin A$ Area formula

$= \frac{1}{2}$ sin   Replace b, c, and A.

$\approx$ [ ] square centimeters   Use a calculator.

Your Turn Find the area of $\triangle ABC$ to the nearest tenth.

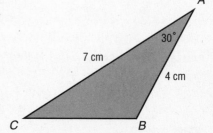

EXAMPLE One Solution

2 **In $\triangle ABC$, $A = 25°$, $a = 13$, and $b = 12$. Determine whether $\triangle ABC$ has *no* solution, *one* solution, or *two* solutions. Then solve $\triangle ABC$.**

Angle A is acute and $a > b$, so one solution exists. Make a sketch and then use the Law of Sines to find B.

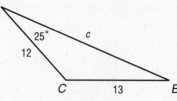

$\dfrac{\sin B}{[\;]} = \dfrac{\sin [\;]}{[\;]}$ Law of Sines

$\sin B = $ [   ]   Multiply by 12.

$\sin B \approx $ [   ]   Use a calculator.

$B \approx $ [   ]   Use the $\sin^{-1}$ function.

FOLDABLES

ORGANIZE IT

On the fourth page of your Trigonometric Functions booklet, write the Law of Sines. Then explain what triangle measures you must know in order to apply the Law of Sines.

The measure of angle C is approximately

$180 -$ [ ] or [ ] .

Use the Law of Sines again to find c.

$$\frac{\sin [\ ]}{c} \approx \frac{\sin [\ ]}{[\ ]} \qquad \text{Law of Sines}$$

$$c \approx [\ ] \qquad \text{or about} \quad [\ ]$$

So, $B \approx$ [ ] , $C \approx$ [ ] , and $c \approx$ [ ] .

Your Turn In $\triangle ABC$, $A = 33°$, $a = 15$, and $b = 10$. Determine whether $\triangle ABC$ has *no* solution, *one* solution, or *two* solutions. Then solve $\triangle ABC$.

[ ]

EXAMPLE Two Solutions

3 In $\triangle ABC$, $A = 25°$, $a = 5$, and $b = 10$. Determine whether $\triangle ABC$ has *no* solution, *one* solution, or *two* solutions. Then solve $\triangle ABC$.

Since angle A is acute, find $b \sin A$ and compare it with a.

$b \sin A =$ [ ]   Replace b and A.

$\approx$ [ ]   Use a calculator.

Since $10 > 5 > 4.23$, there are two possible solutions. Thus, there are two possible triangles to be solved.

CASE 1 Acute Angle B

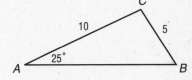

First use the Law of Sines to find B.

$$\frac{\sin B}{10} = \frac{\sin 25°}{5}$$

$$\sin B = [\qquad]$$

$$\sin B \approx [\quad] . \text{ So } B \approx [\quad] .$$

The measure of angle C is approximately

$180 - (25 + 58)$ or ⬚ .

Use the Law of Sines again to find c.

$$\frac{\sin 97°}{c} \approx \frac{\sin 25°}{5}$$

$c \approx$ ⬚

$c \approx$ ⬚

Therefore, $B \approx$ ⬚ , $C \approx$ ⬚ , and $c \approx$ ⬚ .

CASE 2 Obtuse Angle B

To find B, you need to find an obtuse angle whose sin is also 0.8452. To do this, subtract the angle given by your calculator, 58°, from 180°. So B is approximately $180 -$ ⬚ or ⬚ .

The measure of angle C is approximately

$180 -$ ⬚ or ⬚ .

Use the Law of Sines to find c.

$$\frac{\sin 33°}{c} \approx \frac{\sin 25°}{5}$$

$c \approx$ ⬚

$c \approx$ ⬚

Therefore, $B \approx$ ⬚ , $C \approx$ ⬚ , and $c \approx$ ⬚ .

HOMEWORK ASSIGNMENT

Page(s):

Exercises:

Your Turn In $\triangle ABC$, $A = 27°$, $a = 12$, and $b = 20$. Determine whether $\triangle ABC$ has *no* solution, *one* solution, or *two* solutions. Then solve $\triangle ABC$.

Law of Cosines

WHAT YOU'LL LEARN

- Solve problems by using the Law of Cosines.

- Determine whether a triangle can be solved by first using the Law of Sines or the Law of Cosines.

KEY CONCEPT

Law of Cosines Let $\triangle ABC$ be any triangle with a, b, and c representing the measures of sides, and opposite angles with measures A, B, and C, respectively. Then the following equations are true.

$a^2 = b^2 + c^2 - 2bc \cos A$

$b^2 = a^2 + c^2 - 2ac \cos B$

$c^2 = a^2 + b^2 - 2ab \cos C$

EXAMPLE Solve a Triangle Given Two Sides and Included Angle

1 Solve $\triangle ABC$.

Use the Law of Cosines to find c.

$c^2 = a^2 + b^2 - 2ab \cos C$

$c^2 = \boxed{} + \boxed{} - \boxed{}$

$c^2 \approx \boxed{}$ Simplify using a calculator.

$c \approx \boxed{}$ Take the square root of each side.

Next, use the Law of Sines to find the measure of angle A.

$\dfrac{\sin A}{a} = \dfrac{\sin C}{c}$ Law of Sines

$\dfrac{\sin A}{\boxed{}} \approx \dfrac{\sin 73°}{\boxed{}}$ $a = 7$, $C = 73°$, and $c \approx 10.4$

$\sin A \approx \boxed{}$ Multiply each side by 7.

$\sin A \approx \boxed{}$ Use a calculator.

$A \approx \boxed{}$ Use the $\sin^{-1}$ function.

The measure of angle B is approximately $180 - \boxed{}$

or $\boxed{}$. So, $c \approx \boxed{}$, $A \approx \boxed{}$, and $B \approx \boxed{}$.

Your Turn Solve $\triangle ABC$.

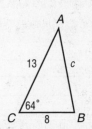

EXAMPLE Solve a Triangle Given Three Sides

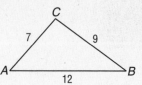

2 Solve △*ABC*.

You are given the measures of three sides.
Use the Law of Cosines to find the
measure of the largest angle first, angle *C*.

$$c^2 = a^2 + b^2 - 2ab \cos C$$

$$\boxed{} = \boxed{} + \boxed{} - \boxed{}$$

$$\boxed{} - 9^2 - \boxed{} = \boxed{}$$

$$\frac{144 - 81 - 49}{\boxed{}} = \boxed{}$$

$$-0.1111 \approx \cos C$$

$$C \approx \boxed{}$$

Use the Law of Sines to find the measure of angle *B*.

$$\frac{\sin B}{b} = \frac{\sin C}{c} \qquad \text{Law of Sines}$$

$$\frac{\sin B}{\boxed{}} \approx \frac{\sin \boxed{}}{\boxed{}} \qquad b = 7,\ C \approx 96°,\ \text{and } c = 12$$

$$\sin B \approx \boxed{} \qquad \text{Multiply each side by 7.}$$

$$\sin B \approx 0.5801 \qquad \text{Use a calculator.}$$

$$B \approx \boxed{} \qquad \text{Use the } \sin^{-1} \text{ function.}$$

The angle of measure *A* is approximately

$180 - \boxed{}$ or $\boxed{}$. So, $A \approx \boxed{}$, $B \approx \boxed{}$,

and $C \approx \boxed{}$.

 Solve △*ABC*.

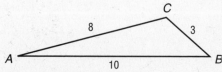

**HOMEWORK
ASSIGNMENT**

Page(s):

Exercises:

© Glencoe/McGraw-Hill

WHAT YOU'LL LEARN

- Define and use the trigonometric functions based on the unit circle.

- Find the exact values of trigonometric functions of angles.

KEY CONCEPT

Definition of Sine and Cosine If the terminal side of an angle θ in standard position intersects the unit circle at $P(x, y)$, then $\cos \theta = x$ and $\sin \theta = y$. Therefore, the coordinates of P can be written as $P(\cos \theta, \sin \theta)$

Periodic Function A function is called periodic if there is a number a such that $f(x) = f(x + a)$ for all x in the domain of the function. The least positive value of a for which $f(x) = f(x + a)$ is called the period of the function.

EXAMPLE Find Sine and Cosine Given Point on Unit Circle

1 Given an angle θ in standard position, if $P\left(\dfrac{\sqrt{7}}{4}, \dfrac{3}{4}\right)$ lies on the terminal side of θ and on the unit circle, find $\sin \theta$ and $\cos \theta$.

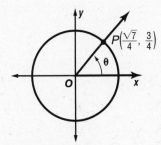

$$P\left(\dfrac{\sqrt{7}}{4}, \dfrac{3}{4}\right) = \boxed{}$$

$\sin \theta = \boxed{}$ and $\cos \theta = \boxed{}$

Your Turn Given an angle θ in standard position, if $P\left(\dfrac{\sqrt{15}}{4}, -\dfrac{1}{4}\right)$ lies on the terminal side of θ and on the unit circle, find $\sin \theta$ and $\cos \theta$.

BUILD YOUR VOCABULARY (page 299)

Every 360° or 2π radians, the sine and cosine functions

$\boxed{}$ their values. Therefore, these functions are

periodic, each having a **period** of $\boxed{}$ or $\boxed{}$ radians.

EXAMPLE Find the Value of a Trigonometric Function

2 Find the exact value of each function.

a. cos 690°

$$\cos 690° = \cos \left(\boxed{} + \boxed{} \right)$$

$$= \cos \boxed{}$$

$$= \boxed{}$$

b. $\sin\left(-\dfrac{3\pi}{4}\right)$

$$\sin\left(-\dfrac{3\pi}{4}\right) = \sin\left(-\dfrac{3\pi}{4}\right) + \boxed{}$$

$$= \sin\left(\boxed{}\right)$$

$$= \boxed{}$$

Your Turn Find the exact value of each function.

a. cos 405°

b. $\sin\left(-\dfrac{\pi}{2}\right)$

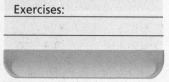
© Glencoe/McGraw-Hill

Inverse Trigonometric Function

WHAT YOU'LL LEARN

- Solve equations by using inverse trigonometric functions.

- Find values of expressions involving trigonometric functions.

BUILD YOUR VOCABULARY (pages 298–299)

The domains of trigonometric functions must be restricted so that their [ ] are functions. The values in these restricted domains are called **principal values**.

The inverse of the Sine function is called the **Arcsine function** and is symbolized by [ ] or [ ].

The definitions of the **Arccosine** and **Arctangent functions** are similar to the definition of the Arcsine function.

KEY CONCEPT

Principal Values of Sine, Cosine, and Tangent

- $y = \text{Sin } x$ if and only if $y = \sin x$ and $-\frac{\pi}{2} \le x \le \frac{\pi}{2}$

- $y = \text{Cos } x$ if and only if $y = \cos x$ and $0 \le x \le \pi$

- $y = \text{Tan } x$ if and only if $y = \tan x$ and $-\frac{\pi}{2} \le x \le \frac{\pi}{2}$

EXAMPLE Solve an Equation

① **Solve Sin $x = \dfrac{\sqrt{2}}{2}$ by finding the value of x to the nearest degree.**

If $\text{Sin } x = \dfrac{\sqrt{2}}{2}$, the x is the least value whose sine is [ ].

So, $x =$ [ ].

Use a calculator to find x.

Keystrokes:

[2nd] [SIN⁻¹] [2nd] [√] 2) ÷ 2) [ENTER]

The value of x is [ ].

Your Turn Solve $\text{Cos } x = \dfrac{\sqrt{3}}{2}$ by finding the value of x to the nearest degree.

EXAMPLE Find a Trigonometric Value

② **Find each value. Write the angle measure in radians. Round to the nearest hundredth.**

a. Arcsin $\dfrac{\sqrt{2}}{2}$

Keystrokes:

| 2nd | [SIN⁻¹] | 2nd | [√] 2 |) | ÷ | 2 |) | ENTER |

Arcsin $\dfrac{\sqrt{2}}{2}$ ≈ ⬚ radian

b. $\tan\left(\text{Cos}^{-1}\dfrac{4}{5}\right)$

Keystrokes:

| 2nd | 2nd | [COS⁻¹] 4 | ÷ | 5 |) | ENTER |

$\tan\left(\cos^{-1}\dfrac{4}{5}\right)$ = ⬚ radian

Your Turn Find each value. Write the angle measures in radians. Round to the nearest hundredth.

a. Arcsin $\left(\dfrac{\sqrt{3}}{2}\right)$ **b.** $\tan\left(\text{Cos}^{-1}\dfrac{2}{3}\right)$

HOMEWORK ASSIGNMENT

Page(s):

Exercises:

STUDY GUIDE

FOLDABLES™	VOCABULARY PUZZLEMAKER	*BUILD YOUR VOCABULARY*
Use your **Chapter 13 Foldable** to help you study for your chapter test.	To make a crossword puzzle, word search, or jumble puzzle of the vocabulary words in Chapter 13, go to: www.glencoe.com/sec/math/t_resources/free/index.php.	You can use your completed **Vocabulary Builder** (pages 298–299) to help you solve the puzzle.

13-1
Right Triangle Trigonometry

1. Find the values of the six trigonometric functions for angle θ.

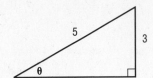

13-2
Angles and Angle Measure

Match each degree measure with the corresponding radian measure on the right.

2. 30°

3. 90°

4. 120°

5. 135°

6. 180°

a. $\dfrac{2\pi}{3}$

b. $\dfrac{\pi}{2}$

c. π

d. $\dfrac{\pi}{6}$

e. $\dfrac{3\pi}{4}$

13-3
Trigonometric Functions of General Angles

7. Find the exact values of the six trigonometric functions of θ if the terminal side of θ in standard position contains the point $(1, -2)$.

13-4
Law of Sines

Determine whether $\triangle ABC$ has _no solution, one solution,_ or _two solutions_. Do not try to solve the triangle.

8. $a = 20$, $A = 30°$, $B = 70°$

9. $c = 12$, $A = 100°$, $a = 30$

10. $C = 27°$, $b = 23.5$, $c = 17.5$

13-5
Law of Cosines

Suppose that you are asked to solve $\triangle ABC$ given the following information about the sides and angles of the triangle. In each case, indicate whether you would begin by using the _Law of Sines_ or the _Law of Cosines_.

11. $a = 8$, $b = 7$, $c = 6$

12. $b = 9.5$, $A = 72°$, $B = 39°$

13. $C = 123°$, $b = 22.95$, $a = 34.35$

14. Solve $\triangle ABC$ for $a = 4$, $b = 5$, and $c = 8$.

13-6

Circular Functions

Tell whether each function is periodic. Write *yes* or *no*.

15. $y = 2x$ [   ]

16. $y = x^2$ [   ]

17. $y = \cos x$ [   ]

18. $y = |x|$ [   ]

Find the period of each function by examining its graph.

19.

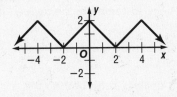

[   ]

20.

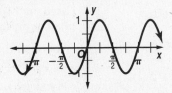

[   ]

Find the exact value of each function.

21. $\sin 765°$

[   ]

22. $\cos\left(-\dfrac{9\pi}{2}\right)$

[   ]

13-7

Inverse Trigonometric Functions

Indicate whether each statement is *true* or *false*.

23. The domain of the function $y = \sin x$ is the set of all real

numbers. [   ]

24. The domain of the function $y = \text{Cos } x$ is $0 \le x \le \pi$. [   ]

25. The range of the function $y = \text{Tan } x$ is $-1 \le y \le 1$. [   ]

Find each value. Round to the nearest hundredth, if necessary.

26. $\text{Cos}^{-1}\dfrac{1}{2}$

[   ]

27. $\sin\left(\text{Tan}^{-1}\dfrac{3}{2}\right)$

[   ]

CHAPTER 13

Checklist

ARE YOU READY FOR THE CHAPTER TEST?

Math Online

Visit **algebra2.com** to access your textbook, more examples, self-check quizzes, and practice tests to help you study the concepts in Chapter 13.

Check the one that applies. Suggestions to help you study are given with each item.

☐ **I completed the review of all or most lessons without using my notes or asking for help.**

- You are probably ready for the Chapter Test.
- You may want to take the Chapter 13 Practice Test on page 75 of your textbook as a final check.

☐ **I used my Foldable or Study Notebook to complete the review of all or most lessons.**

- You should complete the Chapter 13 Study Guide and Review on pages 752–756 of your textbook.
- If you are unsure of any concepts or skills, refer back to the specific lesson(s).
- You may also want to take the Chapter 13 Practice Test on page 757.

☐ **I asked for help from someone else to complete the review of all or most lessons.**

- You should review the examples and concepts in your Study Notebook and Chapter 13 Foldable.
- Then complete the Chapter 13 Study Guide and Review on pages 752–756 of your textbook.
- If you are unsure of any concepts or skills, refer back to the specific lesson(s).
- You may also want to take the Chapter 13 Practice Test on page 757.

Student Signature	Parent/Guardian Signature

Teacher Signature

© Glencoe/McGraw-Hill

Trigonometric Graphs and Identities

 Use the instructions below to make a Foldable to help you organize your notes as you study the chapter. You will see Foldable reminders in the margin this Interactive Study Notebook to help you in taking notes.

Begin with eight sheets of grid paper.

 STEP 1 **Staple**
Staple the stack of grid paper along the top to form a booklet.

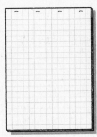

 STEP 2 **Cut and Label**
Cut seven lines from the bottom of the top sheet, six lines from the second sheet, and so on. Label with lesson numbers as shown.

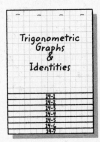

 NOTE-TAKING TIP: When you take notes, write instructions on how to do something presented in each lesson. Then follow your own instructions to check them for accuracy.

BUILD YOUR VOCABULARY

This is an alphabetical list of new vocabulary terms you will learn in
Chapter 14. As you complete the study notes for the chapter, you will see
Build Your Vocabulary reminders to complete each term's definition or
description on these pages. Remember to add the textbook page number
in the second column for reference when you study.

Vocabulary Term	Found on Page	Definition	Description or Example
amplitude [AM-pluh-TOOD]			
double-angle formula			
half-angle formula			
midline			
phase shift [FAYZ]			
trigonometric equation			
trigonometric identity			
vertical shift			

Graphing Trigonometric Functions

© Glencoe/McGraw-Hill

WHAT YOU'LL LEARN

- Graph trigonometric functions.

- Find the amplitude and period of variation of the sine, cosine, and tangent functions.

BUILD YOUR VOCABULARY (page 322)

The **amplitude** of the graph of a periodic function is the [ ] value of half the difference between its [ ] value and its [ ] value.

EXAMPLE Graph Trigonometric Functions

① **Find the amplitude and period of each function. Then graph the function.**

a. $y = \sin \frac{1}{3}\theta$

First, find the amplitude.

$|a| =$ [ ]   The coefficient of $\sin \frac{1}{3}\theta$ is 1.

KEY CONCEPT

Amplitudes and Periods
For functions of the form $y = a \sin b\theta$ and $y = a \cos b\theta$, the amplitude is $|a|$, and the period is $\frac{360°}{|b|}$ or $\frac{2\pi}{|b|}$.
For functions of the form $y = a \tan b\theta$, the amplitude is not defined, and the period is $\frac{180°}{|b|}$ or $\frac{\pi}{|b|}$.

Next, find the period.

$\dfrac{360°}{|b|} =$ [ ] $b = \dfrac{1}{3}$

$= 1080°$ or 6π

Use the amplitude and period to graph the function.

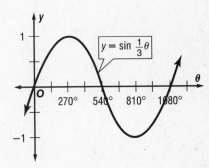

b. $y = \frac{1}{3} \cos \theta$

Amplitude: $|a| =$ [ ]

$=$ [ ]

Period: $\dfrac{360°}{|b|} =$ [ ]

$=$ [ ]

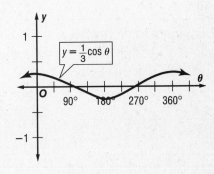

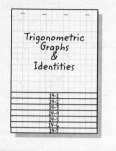

FOLDABLES

ORGANIZE IT

On the page for Lesson 14-1, describe three real-world situations that fluctuate in a regular, periodic pattern.

Trigonometric
Graphs
&
Identities

14-1
14-2
14-3
14-4
14-5
14-6
14-7

c. $y = 2 \sin \frac{1}{4}\theta$.

Amplitude: $|a| = $ ▢

$= $ ▢

Period: $\frac{2\pi}{|b|} = $ ▢

$= $ ▢ or $1440°$

Your Turn Find the amplitude and period for each function. Then graph the function.

a. $y = \sin \frac{1}{2}\theta$

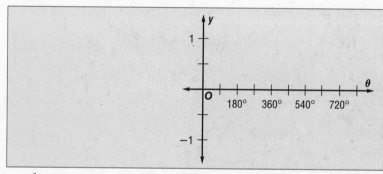

b. $y = \frac{1}{2} \cos \theta$

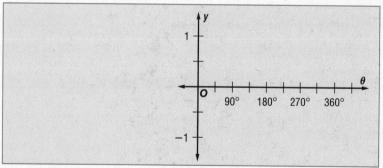

c. $y = 3 \sin \frac{1}{2}\theta$

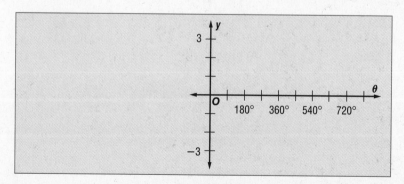

HOMEWORK ASSIGNMENT

Page(s):

Exercises:

Translations of Trigonometric Graphs

WHAT YOU'LL LEARN

- Graph horizontal translations of trigonometric graphs and find phase shifts.

- Graph vertical translations of trigonometric graphs.

KEY CONCEPT

Phase Shift The phase shift of the functions
$y = a \sin b(\theta - h) + k$,
$y = a \cos b(\theta - h) + k$,
and
$y = a \tan b(\theta - h) + k$ is h,
where $b > 0$.

- If $h > 0$, the shift is to the right.

- If $h < 0$, the shift is to the left.

BUILD YOUR VOCABULARY (page 322)

A horizontal translation of a trigonometric function is called a **phase shift**.

EXAMPLE Graph Horizontal Translations

1 State the amplitude, period, and phase shift of $y = 2 \sin (\theta + 20°)$. Then graph the function.

Since $a = \boxed{}$ and $b = \boxed{}$, the amplitude and period of

the function are the same as $y = 2 \cos \theta$. However, $h = \boxed{}$,

so the phase shift is $\boxed{}$. Because $h < 0$ the parent graph is

shifted to the $\boxed{}$.

To graph $y = 2 \sin (\theta + 20°)$, consider the graph of $y = 2 \sin \theta$.

Graph this function and then shift the graph $\boxed{}$ to the $\boxed{}$.

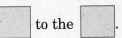
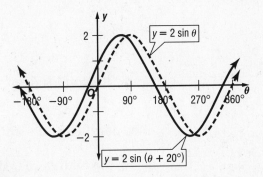

Your Turn State the amplitude, period, and phase shift of $y = \frac{1}{4} \cos \left(\theta - \frac{\pi}{4}\right)$. Then graph the function.

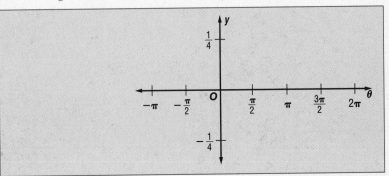

BUILD YOUR VOCABULARY (page 322)

Graphs of trigonometric functions can be translated

[   ] through a **vertical shift**.

A new [   ] axis called the **midline** becomes the

reference line about which the graph [   ].

KEY CONCEPT

Vertical Shift The vertical shift of the functions

$y = a \sin b(\theta - h) + k$,
$y = a \cos b(\theta - h) + k$,
and
$y = a \tan b(\theta - h) + k$ is k.

• If $k > 0$, the shift is up.

• If $k < 0$, the shift is down.

• The midline is $y = k$.

REMEMBER IT

You may find it easier to graph the parent graph in one color, and then each transformation in a different color.

EXAMPLE Graph Vertical Translations

2 **State the vertical shift, equation of the midline, amplitude, and period of $y = \frac{1}{2} \cos \theta + 3$. Then graph the function.**

Vertical shift:

$k = $ [   ], so

the midline is the graph of

$y = $ [   ].

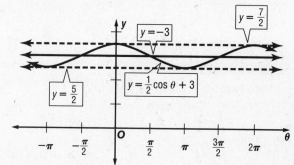

$y = \frac{7}{2}$ $y = -3$ $y = \frac{5}{2}$ $y = \frac{1}{2}\cos\theta + 3$

Amplitude: $|a| = $ [   ]   Period: $\frac{2\pi}{|b|} = $ [   ]

Since the amplitude of the function is [   ], draw dashed

lines parallel to the midline that are [   ] unit above and

below the midline. Then draw the cosine curve.

Your Turn State the vertical shift, equation of the midline, amplitude, and period of $y = 3 \sin \theta - 2$. Then graph the function.

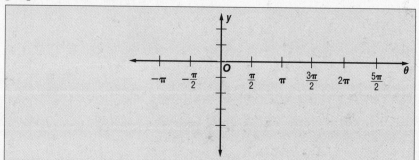

EXAMPLE Graph Transformations

3 State the vertical shift, amplitude, period, and phase shift of $y = 3 \sin\left[2\left(\theta - \dfrac{\pi}{2}\right)\right] + 4$. Then graph the function.

The function is written in the form $y = a \cos[b(\theta - h) + k]$. Identify the values of k, a, b, and h.

STEP 1 The vertical shift is ⬚. Graph the midline

$y = $ ⬚ .

STEP 2 The amplitude is ⬚. Draw dashed lines

⬚ units above and below the midline at

$y = $ ⬚ and $y = $ ⬚ .

STEP 3 The period is ⬚ , so the graph is ⬚ .

Graph $y = 3 \sin 2\theta + 4$ using the midline as a reference.

STEP 4 Shift the graph ⬚ to the ⬚ .

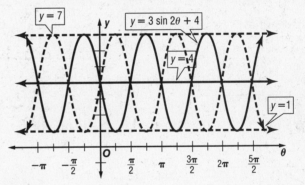

$y = 7$ $y = 3 \sin 2\theta + 4$ $y = 4$ $y = 1$

Your Turn State the vertical shift, amplitude, period, and phase shift of $y = 2 \cos\left[3\left(\theta - \dfrac{\pi}{4}\right)\right] - 2$. Then graph the function.

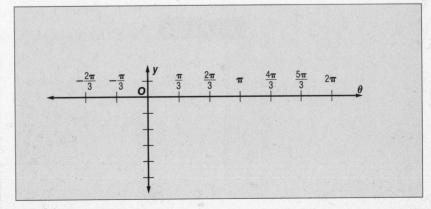

FOLDABLES

ORGANIZE IT

Use the page for Lesson 14-2. Sketch a phase shift and a vertical shift for the sine, cosine, and tangent functions.

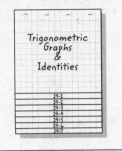

Trigonometric Graphs & Identities

14-1
14-2
14-3
14-4
14-5
14-6
14-7

HOMEWORK ASSIGNMENT

Page(s):

Exercises:

Trigonometric Identities

WHAT YOU'LL LEARN

- Use identities to find trigonometric values.

- Use trigonometric identities to simplify expressions.

BUILD YOUR VOCABULARY (page 320)

A **trigonometric identity** is an equation involving

trigonometric [ ] that is true for [ ] values

for which every expression in the equation is defined.

KEY CONCEPT

Basic Trigonometric Identities

Quotient Identities

$\tan \theta = \dfrac{\sin \theta}{\cos \theta}$

$\cot \theta = \dfrac{\cos \theta}{\sin \theta}$

Reciprocal Identities

$\csc \theta = \dfrac{1}{\sin \theta}$ $\sec \theta = \dfrac{1}{\cos \theta}$

$\cot \theta = \dfrac{1}{\tan \theta}$

Pythagorean Identities

$\cos^2 \theta + \sin^2 \theta = 1$

$\tan^2 \theta + 1 = \sec^2 \theta$

$\cot^2 \theta + 1 = \csc^2 \theta$

EXAMPLE Find a Value of a Trigonometric Function

1 Find $\tan \theta$ if $\sec \theta = -2$ and $180° < \theta < 270°$.

$\tan^2 \theta + 1 = \sec^2 \theta$ — Trigonometric identity

$\tan^2 \theta =$ [ ] — Subtract 1 from each side.

$\tan^2 \theta =$ [ ] — Substitute -2 for $\sec \theta$.

$\tan^2 \theta =$ [ ] — Square -2.

$\tan^2 \theta =$ [ ] — Subtract.

$\tan \theta =$ [ ] — Take the square root of each side.

Since θ is in the third quadrant, $\tan \theta$ is [ ].

Thus, $\tan \theta =$ [ ].

Your Turn Find the value of each expression.

a. $\cos \theta$ if $\sin \theta = \dfrac{1}{3}$ and $0° < \theta < 90°$

b. sec θ if tan $\theta = -2$, and $\dfrac{3\pi}{2} < \theta < 2\pi$

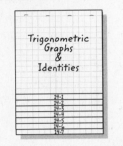
EXAMPLE Simplify an Expression

2 **Simplify sin θ(csc θ − sin θ).**

$\sin \theta(\csc \theta - \sin \theta)$

$= \sin \theta$ [   ] $\csc \theta = \left(\dfrac{1}{\sin \theta}\right)$

$= \sin \theta$ [   ] $- \sin \theta$ [   ]

Distributive Property

$=$ [   ]   Simplify.

$=$ [   ] $1 - \sin^2 \theta = \cos^2 \theta$

Your Turn Simplify tan θ cot θ.

HOMEWORK ASSIGNMENT

Page(s):

Exercises:

Verifying Trigonometric Identities

WHAT YOU'LL LEARN

- Verify trigonometric identities by transforming one side of an equation into the form of the other side.

- Verify trigonometric identities by transforming each side of the equation into the same form.

1 **Verify that csc θ cos θ tan θ = 1 is an identity.**

Transform the left side.

csc θ cos θ tan θ $\overset{?}{=}$ 1   Original equation

 · [   ] · [   ] $\overset{?}{=}$ 1   csc $\theta = \dfrac{1}{\sin \theta}$,

tan $\theta = \dfrac{\sin \theta}{\cos \theta}$

[   ] = 1   Simplify.

2 **Verify that csc θ + sec $\theta = \dfrac{1 + \cot \theta}{\cos \theta}$ is an identity.**

csc θ + sec θ $\overset{?}{=}$ $\dfrac{1 + \cot \theta}{\cos \theta}$ Original equation

WRITE IT

Why do you include a question mark above the equality sign when verifying an identity?

_____

_____

_____

_____

_____

_____

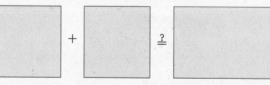 +  $\overset{?}{=}$ [   ]   Express all terms using sine and cosine.

[   ] $\overset{?}{=}$ [   ]   Rewrite using the LCD, sin θ cos θ.

 = [   ]   Simplify the right side.

Your Turn

a. Verify that $\csc^2 \theta \tan^2 \theta = \sec^2 \theta$ is an identity.

ORGANIZE IT

On the page for Lesson 14-4, write your own checklist for verifying trigonometric identities.

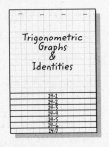

b. Verify that $1 + \dfrac{1}{\cos \theta} = \dfrac{\tan^2 \theta}{\sec \theta - 1}$ is an identity.

HOMEWORK ASSIGNMENT

Page(s):

Exercises:

Sum and Difference of Angles Formulas

WHAT YOU'LL LEARN

- Find values of sine and cosine involving sum and difference formulas.

- Verify identities by using sum and difference formulas.

EXAMPLE Use Sum and Difference of Angles Formulas

① **Find the exact value of each expression.**

a. sin 75°

Use the formula $\sin(\alpha + \beta) = \sin\alpha\cos\beta + \cos\alpha\sin\beta$.

$\sin 75° = \sin(30° + 45°)$ $\alpha = 30°; \beta = 45°$

$= \sin \boxed{} \cos \boxed{} + \cos \boxed{} \sin \boxed{}$

$= \boxed{} \cdot \boxed{} + \boxed{} \cdot \boxed{}$

$= \boxed{} + \boxed{}$

$= \boxed{}$

b. cos (−75°)

Use the formula $\cos(\alpha - \beta) = \cos\alpha\cos\beta + \sin\alpha\sin\beta$.

$\cos(-75°) = \cos(60° - 135°)$ $\alpha = 60°; \beta = 135°$

$= \cos 60° \cos 135° + \sin 60° \sin 135°$

$= \boxed{} \cdot \left(-\dfrac{\sqrt{2}}{2}\right) + \boxed{} \cdot \dfrac{\sqrt{2}}{2}$

$= \boxed{} + \boxed{}$

$= \boxed{}$

KEY CONCEPT

Sum and Difference of Angles Formulas The following identities hold true for all values of α and β.

$\cos(\alpha \pm \beta) =$
$\cos\alpha\cos\beta \mp \sin\alpha\sin\beta$
$\sin(\alpha \pm \beta) =$
$\sin\alpha\cos\beta \pm \cos\alpha\sin\beta$

REMEMBER IT

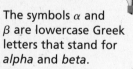

The symbols α and β are lowercase Greek letters that stand for *alpha* and *beta*.

Your Turn **Find the exact value of each expression.**

a. sin 105°

b. cos (−120°)

EXAMPLE Verify Identities

② **Verify that each of the following is an identity.**

a. $\cos (90° - \theta) = \sin \theta$

$$\cos (90° - \theta) \overset{?}{=} \sin \theta$$

$$\underline{\qquad\qquad\qquad\qquad} \overset{?}{=} \sin \theta$$

$$\underline{\qquad\qquad} \overset{?}{=} \sin \theta$$

$$\underline{\qquad} = \sin \theta$$

b. $\cos (180° - \theta) = -\cos \theta$

$$\underline{\qquad\qquad} \overset{?}{=} -\cos \theta$$

$$\underline{\qquad\qquad\qquad} \overset{?}{=} -\cos \theta$$

$$\underline{\qquad\qquad} \overset{?}{=} -\cos \theta$$

$$\underline{\qquad} = -\cos \theta$$

Your Turn Verify that each of the following is an identity.

a. $\cos (90° + \theta) = -\sin \theta$

b. $\sin (180° + \theta) = -\sin \theta$

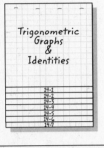
HOMEWORK ASSIGNMENT

Page(s):

Exercises:

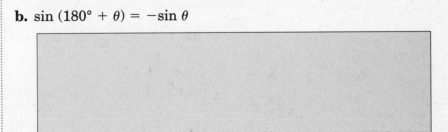

Double-Angle and Half-Angle Formulas

WHAT YOU'LL LEARN

• Find values of sine and cosine involving double-angle formulas.

• Find values of sine and cosine involving half-angle formulas.

KEY CONCEPT

Double-Angles Formulas
The following identities hold true for all values of θ.
$\sin 2\theta = 2 \sin \theta \cos \theta$
$\cos 2\theta = \cos^2 \theta - \sin^2 \theta$
$\cos 2\theta = 1 - 2 \sin^2 \theta$
$\cos 2\theta = 2 \cos^2 \theta - 1$

EXAMPLE Double-Angle Formulas

① **Find the exact value of $\sin 2\theta$ if $\sin \theta = \frac{3}{4}$ and θ is between 0° and 90°.**

Use the identity $\sin 2\theta = 2 \sin \theta \cos \theta$.
First find the value of $\cos \theta$.

$\cos^2 \theta = $ [   ] $\cos^2 \theta + \sin^2 \theta = 1$

$\cos^2 \theta = $ [   ] $\sin \theta = \frac{3}{4}$

$\cos^2 \theta = $ [   ]   Subtract.

$\cos \theta = $ [   ]   Find the square root of each side.

Since θ is in the first quadrant, cosine is [   ].

Thus, $\cos \theta = $ [   ].

Now find $\sin 2\theta$.

$\sin 2\theta = $ [   ]   Double-angle formula

$= $ [   ] $\sin \theta = \frac{3}{4}, \cos \theta = \frac{\sqrt{7}}{4}$

$= $ [   ]   Simplify.

Your Turn Find the value of each expression if $\sin \theta = \frac{1}{4}$ and θ is between 0° and 90°.

a. $\sin 2\theta$ **b.** $\cos 2\theta$

EXAMPLE Half-Angle Formulas

2 Find $\cos \frac{\alpha}{2}$ if $\sin \alpha = \frac{4}{5}$ and α is in the second quadrant.

Since $\cos \frac{\alpha}{2} = \pm\sqrt{\frac{1 + \cos \alpha}{2}}$, we must find $\cos \alpha$ first.

$\cos^2 \alpha = \boxed{}$ $\sin^2 \alpha + \cos^2 \alpha = 1$

$\cos^2 \alpha = \boxed{}$ $\sin \alpha = \frac{4}{5}$

$\cos^2 \alpha = \boxed{}$ Simplify.

$\cos^2 \alpha = \boxed{}$ Take the square root of each side.

Since α is in the second quadrant, $\cos \alpha = \boxed{}$.

$\cos \frac{\alpha}{2} = \pm\sqrt{\frac{1 + \cos \alpha}{2}}$ Half-angle formula

$= \pm\sqrt{\dfrac{1 + \left(-\frac{3}{5}\right)}{2}}$ $\cos \alpha = -\frac{3}{5}$

$= \boxed{}$ Simplify the radicand.

$= \boxed{}$ Rationalize.

$= \boxed{}$ Multiply.

Since α is between 90° and 180°, $\frac{\alpha}{2}$ is between $\boxed{}$ and $\boxed{}$. Thus, $\cos \frac{\alpha}{2}$ is $\boxed{}$ and equals $\boxed{}$.

Your Turn Find $\cos \frac{\alpha}{2}$ if $\sin \alpha = -\frac{9}{41}$ and α is in the fourth quadrant.

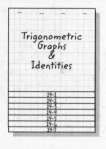
EXAMPLE Evaluate Using Half-Angle Formulas

3 Find the exact value of $\cos \dfrac{9\pi}{8}$ by using the half-angle formulas.

$$\cos \dfrac{9\pi}{8} = \boxed{}$$

$$= \boxed{} \qquad \cos \dfrac{\alpha}{2} = \pm\sqrt{\dfrac{1 + \cos\alpha}{2}}$$

$$= \boxed{} \qquad \cos \dfrac{9\pi}{4} = \dfrac{\sqrt{2}}{2}$$

$$= \boxed{} \qquad \text{Simplify the radicand.}$$

$$= \boxed{} \qquad \text{Simplify the denominator.}$$

Since $\dfrac{9\pi}{8}$ is in the third quadrant, $\cos \dfrac{9\pi}{8}$ is $\boxed{}$.

Thus, $\cos \dfrac{9\pi}{8} = \boxed{}$

Your Turn Find the exact value of each expression by using the half-angle formulas.

a. $\sin 67.5°$

b. $\cos \dfrac{3\pi}{8}$

© Glencoe/McGraw-Hill

Solving Trigonometric Equations

EXAMPLE Solve Equations for a Given Interval

❶ **Find all solutions of $2 \cos^2 \theta - 1 = \sin \theta$ for the interval $0° < \theta \le 360°$.**

$2 \cos^2 \theta - 1 = \sin \theta$	Original equation
$2(1 - \sin^2 \theta) - 1 - \sin \theta = 0$	Solve for 0.
$\boxed{} = 0$	Distributive Property
$\boxed{} = 0$	Simplify.
$2 \sin^2 \theta + \boxed{} - \boxed{} = 0$	Divide each side by -1.
$\boxed{} = 0$	Factor.

Now use the Zero Product Property.

$2 \sin \theta - 1 = 0$ or $\sin \theta + 1 = 0$

$2 \sin \theta = \boxed{}$ $\qquad\qquad$ $\sin \theta = \boxed{}$

$\sin \theta = \boxed{}$ $\qquad\qquad$ $\sin \theta = \boxed{}$

$\theta = \boxed{}$ $\qquad\qquad$ $\theta = \boxed{}$

Your Turn **Find all solutions of each equation for the given interval.**

a. $\sin^2 \theta + \cos 2\theta - \cos \theta = 0; 0° \le \theta < 360°$

b. $\cos \theta + 1 = 0; 0 < \theta \le 2\pi$

EXAMPLE **Solve Trigonometric Equations**

REVIEW IT

Solve $x^2 - x - 6 = 0$ by factoring. *(Lesson 6-3)*

_____

_____

_____

_____

2 Solve $2 \sin \theta \cos \theta = \cos \theta$ for all values of θ if θ is measured in radians.

$2 \sin \theta \cos \theta = \cos \theta$		Original equation
▭ $= 0$		Subtract $\cos \theta$.
▭ $= 0$		Factor.
▭ $= 0$ or ▭ $= 0$		Zero Product Property
▭ $=$ ▭		Solve.

Look at the graph of $y = \sin \theta$ to find solutions of $\sin \theta = \frac{1}{2}$.

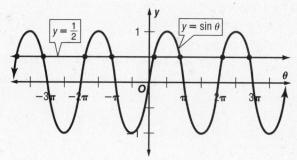

The solutions are $\frac{\pi}{6}, \frac{5\pi}{6}, \frac{13\pi}{6}, \frac{17\pi}{6}$, and so on, and $-\frac{7\pi}{6}, -\frac{11\pi}{6}$, $-\frac{19\pi}{6}, -\frac{23\pi}{6}$, and so on. The only solutions in the interval

0 to 2π are ▭ and ▭ . The period of the sine function

is ▭ radians. So, the solutions can be written as

▭ and ▭ , where k is any

integer. Similarly, the solutions for $\cos \theta = 0$ are

▭ and ▭ .

Your Turn

a. Solve $\cos \theta + 1 = 0$ for all values of θ if θ is measured in radians.

b. Solve $\sin^2 \theta - 1 = \cos^2 \theta$ for all values of θ if θ is measured in degrees.

EXAMPLE Solve Trigonometric Equations Using Identities

③ Solve $\sin \theta \cot \theta = \cos^2 \theta$.

$$\sin \theta \cot \theta = \cos^2 \theta \qquad \text{Original equation}$$

$$\boxed{} = 0 \qquad \cot \theta = \frac{\cos \theta}{\sin \theta}$$

$$\boxed{} = 0 \qquad \text{Multiply.}$$

$$\boxed{} = 0 \qquad \text{Factor.}$$

$$\boxed{} = 0 \quad \text{or} \quad \boxed{} = 0$$

$$\theta = \boxed{} \qquad \boxed{} = \boxed{}$$

$$\theta = \boxed{}$$

Check your solutions. θ equals 0° is extraneous.

The solution is $\boxed{}$.

Your Turn Solve $\sin \theta = \cos \theta$.

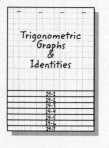
EXAMPLE Determine Whether a Solution Exists

4 Solve $\sin^2 \theta = \dfrac{1}{4} + \cos \theta$.

$$\sin^2 \theta = \frac{1}{4} + \cos \theta \qquad \text{Original equation}$$

$$1 - \cos^2 \theta = \frac{1}{4} + \cos \theta$$

$$\sin^2 \theta = 1 - \cos^2 \theta$$

$$0 = \boxed{} \qquad \begin{matrix}\text{Subtract.}\\\text{Then add.}\end{matrix}$$

$$0 = \boxed{} \qquad \text{Multiply.}$$

$$0 = \boxed{} \qquad \text{Factor.}$$

$$\boxed{} = 0 \ \text{ or } \ \boxed{} = 0$$

$$\cos \theta = \boxed{} \qquad\qquad \cos \theta = \boxed{}$$

No solutions exist. $\qquad\qquad\qquad \theta = \boxed{}$

The solutions are $\boxed{}$ and $\boxed{}$.

Your Turn Solve $\sin 2\theta - \sin \theta = 0$.

STUDY GUIDE

FOLDABLES™	VOCABULARY PUZZLEMAKER	BUILD YOUR VOCABULARY
Use your **Chapter 14 Foldable** to help you study for your chapter test.	To make a crossword puzzle, word search, or jumble puzzle of the vocabulary words in Chapter 14, go to: www.glencoe.com/sec/math/ t_resources/free/index.php	You can use your completed **Vocabulary Builder** (page 322) to help you solve the puzzle.

14-1

Graphing Trigonometric Functions

1. Find the amplitude, if it exists, and period of $y = 2 \sin \frac{1}{4}x$. Then graph the function.

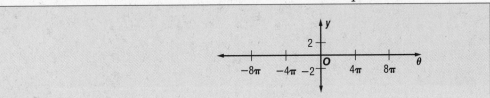

Determine whether each statement is *true* or *false*.

2. The period of a function is the distance between the maximum and minimum points. [   ]

3. The amplitude of the function $y = \sin \theta$ is 2π. [   ]

14-2

Translations of Trigonometric Graphs

Determine whether the graph of each function has an *amplitude change*, *period change*, *phase shift*, or *vertical shift* compared to the graph of the parent function. (More than one of these may apply to each function. Do not actually graph the functions.)

4. $y = 3 \sin\left(\dfrac{\theta + 8\pi}{6}\right)$ [   ]

5. $y = \cos(2\theta + 70°)$ [   ]

6. $y = -4 \cos 3\theta$ [   ]

7. $y = \sec \dfrac{1}{2}\theta - 3$ [   ]

14-3

Trigonometric Identities

8. Find the value of $\sin \theta$, if $\cos \theta = \frac{1}{2}$ and $270° < \theta < 360°$.

Match each expression from the list on the left with an expression from the list on the right that is equal to it for all values for which each expression is defined.

9. $\sec^2 \theta - \tan^2 \theta$

10. $\dfrac{\sin \theta}{\cos \theta}$

11. $\csc \theta$

12. $\dfrac{\cos \theta}{\sin \theta}$

a. $\dfrac{1}{\sin \theta}$

b. $\tan \theta$

c. 1

d. $\cot \theta$

14-4

Verifying Trigonometric Identities

Determine whether each equation is an *identity* or *not an identity*.

13. $\dfrac{1}{\sin^2 \theta} - \dfrac{1}{\tan^2 \theta} = 1$

14. $\dfrac{\sin \theta}{\cos \theta} + \dfrac{\cos \theta}{\sin \theta} = \cos \theta \sin \theta$

15. $\dfrac{\sin^2 \theta}{\cos^2 \theta} + 1 = \sec^2 \theta$

16. $\tan^2 \theta \cos^2 \theta = \dfrac{1}{\csc^2 \theta}$

17. Verify that $\csc \theta (\cos \theta + \sin \theta) = \cot \theta + 1$.

14-5
Sum and Difference of Angles Formulas

Find the exact value of each expression.

18. $\sin 165°$

19. $\cos (-210°)$

14-6
Double-Angle and Half-Angle Formulas

Find the exact value for each.

20. $\cos \frac{\theta}{2}$, if $\cos \theta = \frac{2}{3}$; $270° < \theta < 360°$

21. $\sin \theta$, if $\cos 2\theta = \frac{4}{5}$; $90° < \theta < 180°$

Match each expression from the list on the left with *all* expressions from the list on the right that are equal to it for all values of β.

22. $\sin \frac{\beta}{2}$

23. $\cos 2\beta$

24. $\cos \frac{\beta}{2}$

25. $\sin 2\beta$

a. $2 \sin \beta \cos \beta$

b. $1 - 2 \sin^2 \beta$

c. $\cos^2 \beta - \sin^2 \beta$

d. $\pm\sqrt{\dfrac{1 + \cos \beta}{2}}$

e. $\pm\sqrt{\dfrac{1 - \cos \beta}{2}}$

14-7
Solving Trigonometric Equations

26. Find all solutions for $2 \cot \theta = -2$; $0° \le \theta < 360°$.

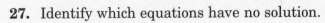

27. Identify which equations have no solution.

a. $\sin \theta = 1$

b. $\tan \theta = 0.001$

c. $\sec \theta = \frac{1}{2}$

d. $\csc \theta = -3$

e. $\cos \theta = 1.01$

f. $\cos \theta + 2 = -1$

ARE YOU READY FOR THE CHAPTER TEST?

Math Online

Visit **algebra2.com** to access your textbook, more examples, self-check quizzes, and practice tests to help you study the concepts in Chapter 14.

Check the one that applies. Suggestions to help you study are given with each item.

☐ **I completed the review of all or most lessons without using my notes or asking for help.**

- You are probably ready for the Chapter Test.
- You may want to take the Chapter 14 Practice Test on page 809 of your textbook as a final check.

☐ **I used my Foldable or Study Notebook to complete the review of all or most lessons.**

- You should complete the Chapter 14 Study Guide and Review on pages 805–808 of your textbook.
- If you are unsure of any concepts or skills, refer back to the specific lesson(s).
- You may also want to take the Chapter 14 Practice Test on page 809 of your textbook.

☐ **I asked for help from someone else to complete the review of all or most lessons.**

- You should review the examples and concepts in your Study Notebook and Chapter 14 Foldable.
- Then complete the Chapter 14 Study Guide and Review on pages 805–808 of your textbook.
- If you are unsure of any concepts or skills, refer back to the specific lesson(s).
- You may also want to take the Chapter 14 Practice Test on page 809 of your textbook.

Student Signature	Parent/Guardian Signature

Teacher Signature